Science and Fiction

Science and Fiction – A Springer Series

This collection of entertaining and thought-provoking books will appeal equally to science buffs, scientists and science-fiction fans. It was born out of the recognition that scientific discovery and the creation of plausible fictional scenarios are often two sides of the same coin. Each relies on an understanding of the way the world works, coupled with the imaginative ability to invent new or alternative explanations—and even other worlds. Authored by practicing scientists as well as writers of hard science fiction, these books explore and exploit the borderlands between accepted science and its fictional counterpart. Uncovering mutual influences, promoting fruitful interaction, narrating and analyzing fictional scenarios, together they serve as a reaction vessel for inspired new ideas in science, technology, and beyond.

Whether fiction, fact, or forever undecidable: the Springer Series "Science and Fiction" intends to go where no one has gone before!

Its largely non-technical books take several different approaches. Journey with their authors as they

- Indulge in science speculation – describing intriguing, plausible yet unproven ideas;
- Exploit science fiction for educational purposes and as a means of promoting critical thinking;
- Explore the interplay of science and science fiction – throughout the history of the genre and looking ahead;
- Delve into related topics including, but not limited to: science as a creative process, the limits of science, interplay of literature and knowledge;

Readers can look forward to a broad range of topics, as intriguing as they are important. Here just a few by way of illustration:

- Time travel, superluminal travel, wormholes, teleportation
- Extraterrestrial intelligence and alien civilizations
- Artificial intelligence, planetary brains, the universe as a computer, virtual worlds
- Non-anthropocentric viewpoints
- Synthetic biology, genetic engineering, developing nanotechnologies
- Eco/infrastructure/meteorite-impact disaster scenarios
- Future scenarios, transhumanism, posthumanism, intelligence explosion
- Consciousness and mind manipulation

Andrew May

How Space Physics Really Works

Lessons from Well-Constructed Science Fiction

Andrew May
Crewkerne, UK

ISSN 2197-1188 ISSN 2197-1196 (electronic)
Science and Fiction
ISBN 978-3-031-33949-3 ISBN 978-3-031-33950-9 (eBook)
https://doi.org/10.1007/978-3-031-33950-9

This Springer imprint is published by the registered company Springer Nature Switzerland AG
The registered company address is: Gewerbestrasse 11, 6330 Cham, Switzerland

Contents

1

Physics in Science Fiction

Mass-market science fiction—the kind we see in movies and TV shows—is notoriously inaccurate when it comes to portraying the physics of outer space. Yet the science in question isn't particularly advanced. The basic principles that govern how a spaceship would move in interplanetary space, or how objects would behave on a planet other than the Earth, have been known for centuries. Fortunately, there has always been a small but dedicated minority of science fiction writers who have taken the trouble to portray physics as accurately as possible in their stories. In this chapter, we take a look at how a few of them—from Jules Verne and Arthur C. Clarke to Larry Niven and Andy Weir—have employed their knowledge of physics to make their stories both more realistic and more enjoyable.

1.1 Celestial Mechanics

More than 60 years after the first humans flew in space, the science fiction (SF) adventures that we see on movie and TV screens still get the basic physics of space travel consistently wrong. Ironically, the conventions of the genre are now so well established that, if physical effects were shown working the way they do in the real world, many viewers would perceive it as an error. As the excellent *TV Tropes* website says:

> Thanks largely to science fiction, space is probably one of the most inaccurately portrayed things in modern media, to the extent that complete falsehoods are widely accepted as fact [1].

A. May, *How Space Physics Really Works*, Science and Fiction,
https://doi.org/10.1007/978-3-031-33950-9_1

The site then goes on to list dozens of examples, including the following:

- All planets are portrayed as having the same surface gravity despite differences in size and mass;
- After a hole is opened in a space ship's outer structure, it has about the same effects as a nearby tornado;
- Spaceships can travel faster than light;
- Objects enter the atmosphere without burning up;
- Turning your engines off in space causes you to slow to a stop.

It's not the purpose of this book to criticize or debunk all the many scientific errors found in mass-media SF. That would be missing the point, because these tales aren't intended to be educational; they're simply fantasy adventures that have their own internal logic and are only meant to be enjoyed by audiences in those terms. Most episodes of *Star Trek*, for example, would be ruined if they had to obey all the laws of physics as they're understood today.

Instead, we'll take a close look at how space *really* works, with the aid of numerous examples taken from written works of SF that actually get their facts right. As a counterbalance to the familiar Hollywood fare, a few authors—from Arthur C. Clarke to Andy Weir (see Fig. 1.1)—have taken positive pleasure in portraying the physics of space as accurately as possible. This has a double benefit, in making the fiction more robust and enjoyable for readers who have a physics background, while making the physics more fun to learn

Fig. 1.1 Andy Weir is among the relatively small number of SF authors who take the trouble to portray the physics of space travel correctly (NASA image)

for those who don't. In fact, outer space is the perfect physics classroom—as another SF author in this category, Larry Niven, has one of his characters observe, when he describes space as "a place to learn physics where you can watch it happen" [2].

We're not going to look at all those advanced technologies—such as beam weapons, cloaking devices, warp drives and teleportation—that are found in SF and may have some tenuous grounding in the more speculative areas of modern physics. There are already plenty of other books that cover such topics, such as *Ten Billion Tomorrows* by Brian Clegg [3] and *All the Wonder that Would Be* by Stephen Webb [4]—the latter in the same "Science and Fiction" series as the present book.

Instead, we're going to focus on well-established areas of physics that—despite the fact that so few movie producers seem to understand them—are really very basic, in the sense that students commonly encounter them in the final years of high-school or the first year of undergraduate studies. This isn't cutting-edge science by any means, with much of it going back at least to the time of Sir Isaac Newton in the seventeenth century. It was Newton who laid down the essentials of what is now known as "classical mechanics" in his book *Philosophiae Naturalis Principia Mathematica* (Mathematical Principles of Natural Philosophy), first published in 1687. This was the work that Arthur C. Clarke, in his novel *2010: Odyssey Two*, uncompromisingly referred to as "the greatest single achievement of the human intellect" [5].

Newton's ideas were expanded by later scientists such as the French polymath Pierre-Simon de Laplace, who consolidated its implications for the way the Solar System works in his evocatively titled magnum opus *Mécanique Céleste* (Celestial Mechanics), published in 5 volumes between 1799 and 1825. However, it was another Frenchman, Jules Verne in the latter half of the nineteenth century, who brought the "mechanics" of space to a much wider audience—and he did so through the medium of science fiction.

1.2 In the Footsteps of Jules Verne

Jules Verne's novel *De la Terre à la Lune* (From the Earth to the Moon) was published in 1865, with its sequel, *Autour de la Lune* (Around the Moon) following in 1870. By that time, the idea of a fictional voyage to the Moon was by no means new, but Verne added a twist of his own that was a genuine first in the world of imaginative fiction: he went out of his way to make the story as consistent as he could with the scientific knowledge of time. He wasn't subtle about it, either, bombarding the reader with a host of technical data and scholarly references throughout both books.

Sadly, this isn't something that's always appreciated by English-language readers, for the simple reason that Verne's early translators seem to have been arts-educated types who completely failed to understand Verne's cutting-edge scientific arguments. So, for example, they rarely made the correct conversions between the metric units used by Verne and the imperial units in use in Britain and America at the time. They mistranslated many technical terms, because—presumably—they weren't aware of the equivalent technical terms in English. And they glossed over Verne's meticulous accounts of the latest academic theories and discoveries, perhaps because so many of them were made by scientists they hadn't heard of.

On the other hand, reading the French original—or a faithful modern translation of it—shows just how scientifically literate Verne was. The second book in particular, *Around the Moon*, can almost be considered a nineteenth century equivalent of Andy Weir's 2014 novel *The Martian*, with physical science so prominent that it's virtually the book's chief protagonist. Like Weir, Verne doesn't merely understand the science—he positively *loves* it.

Also like Weir, Verne wasn't afraid to twist the scientific facts here and there, just so he would have a story to tell in the first place. Weir, for example, has openly acknowledged that the violent Martian storm that starts his novel would never actually happen on Mars [6]. In the same way, Verne knew of no realistic way, in the context of nineteenth century science, to get his voyagers into space and en route to the Moon. So he chose to launch them in a shell-like projectile blasted out of a giant gun—one of the few really egregious impossibilities in his narrative.

The design and constructive of this fanciful "space gun" is the main focus of the first book, *From the Earth to the Moon*. By the time we get to *Around the Moon*—and our protagonists are in space and on their way to the Moon—Verne's science straightens out and becomes as textbook-accurate (with a few excusable exceptions) as anything you'll find anywhere in SF.

As mentioned earlier, the physics of space travel is, for the most part, pretty basic stuff—but "basic" isn't the same as "common sense". Take what may be the simplest of all physics principles, for example: Newton's first law of motion. This states that an object will continue to move at a constant speed in a straight line unless it is acted upon by a force. This is essentially the principle of inertia: it takes an applied force to start a material object moving, and another force, in the opposite direction, to stop it. This sounds simple enough, yet it's not at all intuitive for people born and brought up on Earth. In fact, to anyone who hasn't had Newton's first law drummed into them from an early age, it sounds downright wrong. For the majority of SF movie viewers, the natural tendency of any object is to slow down and stop. That's because there are so many retarding forces here on Earth, such as friction and air resistance, that we've come to assume they exist everywhere in the universe.

In truth, however, those forces are virtually non-existent in outer space, and it's in this environment that Newton's first law really comes into its own. Jules Verne understood this perfectly, as he demonstrates by the fact that his space projectile—after receiving its initial impetus at launch—travels unimpeded once it's beyond the Earth's atmosphere.

There's one very striking, not to say macabre, scene in *Around the Moon* that demonstrates Newton's first law very clearly. The three crewmembers are accompanied on their journey by a pair of dogs, one of which unfortunately dies early in the flight. By analogy with a traditional burial at sea, they decide to eject its body from the projectile. But unlike the maritime analogy, the dog's carcass doesn't fall behind as the projectile continues on its way, but follows alongside all the way to the Moon (see Fig. 1.2).

As Verne recounts in Chap. 6 of *Around the Moon*:

> When Michel Ardan came down, he went to the side scuttle; and suddenly they heard an exclamation of surprise.
>
> "What is it?" asked Barbicane.
>
> The president approached the window, and saw a sort of flattened sack floating some yards from the projectile. This object seemed as motionless as the projectile…
>
> "What astonishes me," said Nicholl, "is that the specific weight of the body, which is certainly less than that of the projectile, allows it to keep so perfectly on a level with it."
>
> "Nicholl," replied Barbicane, after a moment's reflection, "I do not know what the object it, but I do know why it maintains our level."
>
> And why?
>
> Because we are floating in space, my dear captain, and in space bodies fall or move (which is the same thing) with equal speed whatever be their weight or form…

Fig. 1.2 An illustration from Jules Verne's *Around the Moon*, showing—correctly—that the body of the dead dog follows the same trajectory as the projectile itself (public domain image)

> "Just so," said Nicholl, "and everything we throw out of the projectile will accompany it until it reaches the Moon…"
>
> "Ah!" exclaimed Michel, in a loud voice.
>
> "What is the matter?" asked Nicholl.
>
> I know, I guess, what this pretended meteor is! It is no asteroid which is accompanying us … it is our unfortunate dog! [7]

It's only when the projectile's retro-rockets are fired in preparation for its return to Earth that its trajectory diverges from that of the dead dog. This is exactly what happens to ejected objects in real space missions, but it was very prescient of Verne to anticipate it in 1870.

Various writers of serious SF since Verne's time have added realistic details to their stories in a similar way. Much later, but still before the first real-world spaceflight, Isaac Asimov's 1952 short story "The Martian Way" described how a spacewalking astronaut would seem to float beside his spaceship:

> Because your motion was equivalent to that of the ship, it seemed as motionless below you as though it had been painted against an impossible background, while the cable between you hung in coils that had no reason to straighten out [8].

This is all due to that first law of Newton's—that objects continue to travel on the same path unless they're acted on by a force. So two objects having the same initial motion will stay together if neither of them is subjected to a force. On the other hand, if an object is ejected from a spaceship that is being thrust forwards by a rocket motor, then the unpowered object that is jettisoned really will fall behind, just like an object dumped from a moving ship on Earth. That's what happens in Andy Weir's 2021 novel *Project Hail Mary*, when the protagonist has to eject the bodies of a number of dead crewmates. As Weir says, "with the constantly accelerating ship, the body simply falls away." [9].

Another physical effect that Jules Verne describes accurately is the reduced gravitational pull that a person would feel on the lunar surface. Of course, that's not something that anyone had experienced first-hand in 1870, but it's a natural consequence of the way gravity works according to Newtonian theory (a subject we'll look at in more detail in the next chapter). Although Verne's characters don't actually land on the Moon, it was their original intention to do so—and in anticipation of this, they discuss what being there would feel like:

> If no orb exists from whence all laws of weight are banished, you are at least going to visit one where it is much less than on the Earth.
>
> The Moon?
>
> Yes, the Moon, on whose surface objects weigh 6 times less than on the Earth, a phenomenon easy to prove.
>
> "And we shall feel it?" asked Michel.

> Evidently, as 200 pounds will only weigh 30 pounds on the surface of the Moon. [10]

Again, this is something that is far from intuitive. Because the Earth's gravity is essentially the same everywhere on the planet, it's difficult for us to imagine that the same object would have a different weight on the Moon. But "weight" is a measure of the pull of gravity, and the latter is a lot weaker on the Moon that it is on Earth. By the same token, objects fall much more slowly in lunar gravity than they do here on Earth. Like the difference in weight, the difference in fall time is something that can readily be calculated using the equations that describe the force of gravity.

A relative newcomer to writing SF is former astronaut Chris Hadfield (see Fig. 1.3), whose alternate-history thriller *The Apollo Murders* was published in 2021. Not surprisingly, given his background, Hadfield is another writer who likes to get his scientific facts right. Here's what one of his characters says regarding a hole discovered on the Moon that's almost a hundred metres deep:

> If you jumped in, with the Moon's low gravity, you'd slowly fall for 11 seconds. But you'd hit the bottom at 40 miles per hour… I did the math [11].

The laws of gravity and motion that we've encountered so far originated with Isaac Newton, but other, even more basic, physical principles were old even in Newton's time. Chief among these is Archimedes' principle, which Wikipedia dates to circa 246 BCE:

Fig. 1.3 After Canadian astronaut Chris Hadfield retired from space travel, he tried his hand at writing fiction with *The Apollo Murders* (NASA image)

> Any object, totally or partially immersed in a fluid or liquid, is buoyed up by a force equal to the weight of the fluid displaced by the object [12].

The most obvious consequence of this is that, if an object weighs less than the same volume of water, then it will float. This is another bit of physics that Verne uses in *Around the Moon*—right at the end, when the travellers return to Earth. Just like the real-world Apollo missions, the flight ends with the projectile splashing down in the Pacific Ocean. This could have been a disaster if the capsule had gone on to sink to the bottom, but fortunately—to paraphrase one of Verne's characters—the projectile only weighs 19,250 pounds, but displaces 56,000 pounds, and consequently it floats [13].

At first sight, Archimedes' principle might not appear to have much potential in the context of SF, but a few authors since Verne's time have made good use of it. Take Arthur C. Clarke's novella "A Meeting with Medusa" from 1972, for example. It's about a mission that descends into the upper atmosphere of Jupiter using a giant balloon—and the way balloons work all comes down to Archimedes' principle.

Before we look at Clarke's Jupiter balloon, let's think about the more familiar case of a balloon in the Earth's atmosphere. The latter is essentially an 80:20 mix of nitrogen and oxygen, with a density that steadily decreases with increasing altitude. If you fill a large balloon with a gas that's lighter than oxygen and nitrogen—such as hydrogen, which is the lightest gas of all—then by Archimedes' principle it will rise until the density of the surrounding air has dropped to the same density as the hydrogen inside the balloon. At this point—which may be as high as 40 kilometres or more—the balloon stops rising.

If we want to apply the same principle to Jupiter, we hit an immediate problem—because Jupiter's atmosphere itself is largely composed of hydrogen. The only way we can make a balloon that will float in such an atmosphere is to heat the hydrogen inside it, in the same way that a hot-air balloon will float in the Earth's atmosphere despite having the same mixture of gases inside it. As Clarke says in his story: "Only one kind of balloon will work in an atmosphere of hydrogen, which is the lightest of all gases—and that is a hot-hydrogen balloon." [14].

In the Earth's atmosphere, a hydrogen balloon needs to be made from very lightweight material in order to float effectively. On the other hand, on a planet such as Venus that has a much thicker atmosphere, the "balloon" could be a much sturdier structure—even a whole spaceship, such as the one Larry Niven describes in his 1965 story "Becalmed in Hell":

> Our ship hung below the Earth-to-Venus hydrogen fuel tank, 20 miles up and all but motionless in the syrupy air. The tank, nearly empty now, made an excellent blimp. It would keep us aloft as long as the internal pressure matched the external [15].

Returning one last time to Jules Verne's *Around the Moon*, perhaps the most impressive episode—from a science-education perspective—comes in its fourth chapter, dauntingly titled "A Lesson in Algebra". In it, the characters discuss the speed they had to achieve in order to travel all the way from the Earth to the Moon. The episode is impressive for three different reasons. First, it wouldn't have been at all obvious to Verne's original readers—or to many general readers today, for that matter—that the hardest part of getting from A to B in space is a question of attaining the correct starting speed. Secondly, Verne explains the physics behind this—correctly, even by today's standards—in such meticulous detail that it takes up virtually the whole chapter. Thirdly, breaking the tacit "no mathematics" rule that pervades almost all writing aimed at non-scientists, his explanation is accompanied by an algebraic equation (see Fig. 1.4).

We can imagine that for readers in Verne's time, this equation might just as well have been meaningless hieroglyphics—and even today many readers may simply skip over it as the kind of pseudoscientific "technobabble" common in present-day SF. Yet Verne's equation is based on a perfectly real physical principle known as the "conservation of energy"—something we'll return to in more detail later in this chapter. His characters use this principle to derive their equation, and then replace the algebraic symbols with appropriate numbers to derive the desired velocity. That's exactly how real physicists work.

— J'aurai donc alors : *x* égale neuf dixièmes de *d*, et *v* égale zéro, et ma formule deviendra... »

Barbicane écrivit rapidement sur le papier :

$$v_0^2 = 2gr\left\{1 - \frac{10r}{9d} - \frac{1}{81}\left(\frac{10r}{d} - \frac{r}{d-r}\right)\right\}$$

Nicholl lut d'un œil avide.

« C'est cela ! c'est cela! s'écria-t-il.

— Est-ce clair? demanda Barbicane.

— C'est écrit en lettres de feu ! répondit Nicholl.

Fig. 1.4 Excerpt from an early French edition of Jules Verne's *Autour de la Lune*, complete with a mathematically correct equation (Internet Archive)

Verne does make a few simplifications, such as ignoring the relative orbital movements of the Earth and Moon while the projectile is in flight—but real-world scientists often make similar simplifications if they just want an approximate "back-of-the-envelope" estimate. When allowance is made for these simplifying assumptions, Verne's mathematical reasoning and numerical calculations are perfectly correct. We won't go through the whole argument here, but just take a quick canter through it to see how it hangs together.

The starting point is a concept many people may have encountered at school. When you're looking at, for example, the motion of a ball thrown up in the air, it has two types of energy which always add to the same amount. First there's the energy due to its motion through the air—called its kinetic energy (KE)—which is given numerically by half its mass multiplied by the square of its velocity: $\frac{1}{2}mv^2$. Then there's the ball's gravitational potential energy (PE) due to its height above the Earth's surface. This is given by its mass multiplied by the acceleration due to gravity and the height of the ball: mgh. The conservation of energy means that, if the ball is initially thrown upwards from the ground at speed v_0, then it will reach a peak (where its speed is zero) at height h given by:

$$v_0^2 = 2gh$$

(We've cancelled out the m which appears on both sides, and multiplied through by 2 to remove the fraction). This is already looking a little like Verne's equation, which is based on a similar logic applied to the problem of Earth-Moon travel. In the latter case, the aim is to make the "ball" (i.e. space projectile) reach zero speed just beyond the point where the Earth's gravity and the Moon's gravity are pulling with equal force in opposite directions. Beyond this point, the projectile will simply fall towards the Moon instead of falling back to Earth. All the extra terms in Verne's equation—compared with our simplified $v^2 = 2gh$—come from the following three considerations:

- Over space-travel distances, the PE due to the Earth's gravity is a more complicated function than simply mgh;
- The competing pull of the Moon's gravity also has to be taken into account;
- The best reference point isn't the Earth's surface (i.e. ground level) but its centre, with r being the radius of the Earth and d the distance to the Moon.

By the standards of celestial mechanics, Verne's calculation is a very basic one, but by the standards of science fiction—whether in his time or our own—it's an awesomely complicated one. A present-day SF author might encapsulate the physics in a brief statement like "the speed we need is slightly less than the Earth's escape velocity, which we can look up on the internet"—but they would never dream of taking the reader through the actual calculation. Yet Verne did just that—and, within his simplifying assumptions, he did it correctly [16].

1.3 Calculations in Space

Perhaps the most important characteristic of the physical sciences, for anyone involved in them, is that they are fundamentally mathematical in nature. This isn't something that always comes across in popular science writing—let alone science fiction—for the very good reason that it would frighten off the majority of readers. But the fact is that most scientists and engineers automatically think in mathematical terms, just like Verne's space travellers when they were discussing the practicalities of a trip to the Moon.

In a way, it's strange that so many people are frightened by the word "mathematics", because at some level they use it as regularly and unconsciously as professional physicists do. If you have to travel 100 kilometres, and you know that traffic conditions will only allow an average speed of 50 kilometres per hour, then simple common sense says that the journey will take around 2 h. But that "common sense" is really mathematics. If v is average speed, d is distance and t is the time taken, then you're tacitly using the equation $t = d/v$ and substituting in the numerical values $d = 100$ and $v = 50$.

Ultimately, that's all there is to the mathematics of physics: textbook equations with known numerical values substituted into them. But besides everyday numerical quantities like time and distance, physicists quantify other properties such as force and energy. Such words are often used rather loosely in ordinary speech, but they have precise mathematical definitions when they're used by scientists.

The result is that, even in quite mundane situations, people with an ingrained physics background often look at things in a more mathematical way than another person might. Take the protagonist of Andy Weir's novel *Project Hail Mary*, for example. He's a high-school physics teacher, and so he thinks of something as simple and everyday as a pendulum in a way that is almost the verbal equivalent of an equation:

> Cool thing about pendulums: The time it takes for one to swing forward and backward—the period—won't change, no matter how wide it swings. If it's got

a lot of energy, it'll swing farther and faster, but the period will still be the same. This is what mechanical clocks take advantage of to keep time. That period ends up being driven by two things, and two things only: the length of the pendulum and gravity [17].

A pendulum is one of the very simplest mechanical systems, but even far more complicated ones can similarly be described in mathematical terms. Physicists and engineers have, for example, developed equations that describe the motion of cars, ships and aircraft here on Earth. Fortunately, we don't need to master those equations in order to learn how to handle such vehicles—not because physics is irrelevant to the situation, but because we very quickly develop an intuitive understanding of how physics works in our own familiar environment. When an inexperienced driver flips their car over if they turn a corner too fast, or loses traction on an icy bend, they're learning a physics lesson whether they realize it or not.

Turning to motion in outer space, on the other hand, this is no longer true. The equations of rocket science and orbital dynamics simply aren't intuitive in the same way as those applying to a car driving along a highway. So real-world astronauts need to know the mathematics of their subject much more thoroughly than, say, an airline pilot needs to understand the finer points of aerodynamic theory. As we'll see in the third chapter, on orbital dynamics, one of the most famous of the early NASA astronauts—Buzz Aldrin, the second person to walk on the Moon—actually started his career in astronautics by writing a PhD thesis on the mathematical theory of space rendezvous.

You might think this need for a mathematical appreciation of outer space only applied to the earliest astronauts, but actually it's going to apply to anyone who was born and raised on Earth—in other words, virtually everyone for a very long time to come. This is a fact of life that rarely comes across in SF, although Arthur C. Clarke did make passing mention of it in his early novel *The Sands of Mars*, way back in 1951. The protagonist of that novel, as it happens, is an SF author himself, on his first interplanetary flight. To help him understand what is going on, one of the crewmembers recommends him to read a book called *The Elements of Astronautics*. In Clarke's words, the protagonist duly …

… turned over the sheets with an interest that rapidly evaporated as he saw how quickly the proportion of words per page diminished. He finally gave up halfway through the book after coming across a page where the only sentence was "Substituting for the value of perihelion distance from Equation 15.3, we obtain …" All else was mathematics [18].

One person who certainly wasn't put off by mathematical equations was Sir Fred Hoyle, a professor of astrophysics at Cambridge University who also wrote the occasional SF story in his spare time. One of his best-known novels—and a masterpiece of "scientific" SF on a par with *Around the Moon* before it and *The Martian* after it—is *The Black Cloud*, from 1957. It's entirely set on Earth, albeit an Earth beset by a threat from outer space in the form of an interstellar gas cloud heading more or less straight for the Sun. Early in the novel, Hoyle shows how a group of astronomers use a combination of measured data and mathematical equations to estimate the arrival time of the cloud [19].

Amusingly, rather than simplifying the calculation for the sake of his readers, Hoyle actually makes it look more complicated than it needs to be. He even includes a footnote containing a differential equation—from the advanced branch of mathematics known as calculus—when a much more straightforward algebraic equation would have sufficed. In fact, Hoyle's calculation really boils down to the equation we used right at the start of this section: time equals distance divided by speed. We used it to estimate the duration of a car journey, but it equally well applies to the motion of a cloud of gas through interstellar space.

Scientists and engineers often refer to this kind of quick calculation as a "back-of-the-envelope" estimate. There's another good example in Arthur C. Clarke's 1993 novel *The Hammer of God*. This too deals with a cosmic threat approaching Earth, this time in the form of an asteroid that appears to be on a collision course for our planet. Clarke invokes the "half mass times velocity squared" formula for kinetic energy to estimate the destructive potential of the asteroid:

> The mass of Kali was know to within one per cent, and the velocity it would have when meeting Earth was known to 12 decimal places. Any schoolboy could work out the resulting $\frac{1}{2}mv^2$ of energy, and convert it into megatons of explosion [20].

(Although he was a great writer in so many other respects, it's sadly typical of Clarke that he wrote "schoolboy" here rather than "schoolchild"; throughout his career he seems to have had difficulty acknowledging the existence of the female sex).

The point Clarke is making is that it's the speed of an incoming object, even more than its mass, that determines how much destruction it will wreak. Hollywood movies often feel the need to exaggerate the physical size of impactors, because they don't trust their audiences to accept that a relatively small one could cause global levels of devastation. The extreme example of this is *Armageddon* (1998), in which an incoming asteroid is described as "the size of Texas"—in other words, about 1200 km across. Yet there simply aren't any asteroids that big. The largest object in the Asteroid Belt is Ceres, which is just under 1000 km in diameter—and classified as a dwarf planet, not an asteroid [21].

In the real world, even a fairly small asteroid could cause worldwide chaos if it hit the planet squarely at a very high relative velocity (see Fig. 1.5). The object that's believe to have caused the extinction of the dinosaurs—along with many other species—66 million years ago, for example, was only about 10 kilometres in diameter. That's the size of Dallas, not the size of Texas.

Fig. 1.5 A relatively small asteroid could cause widespread devastation on Earth if it collided at high velocity, and hence transferred a large amount of kinetic energy (Wikimedia Commons, CC-BY-2.0)

Another novel that goes into even more detail on the subject is *Lucifer's Hammer* (1977), by Larry Niven and Jerry Pournelle. In this case the anticipated impactor is a comet, the head of which is just a cubic mile in volume and made of low density, loosely packed ice. This doesn't sound all that dangerous, but a group of characters work through the calculations step by step all the same. Knowing the volume of the comet and the density of the material it's composed of, they multiply those two numbers together to estimate its mass.

The other thing they need to know is the speed the comet is going to hit at, which requires a basic knowledge of how orbits work in the Solar System—something we'll learn more about in the next two chapters, on gravity and orbital dynamics. Suffice to say the speed of a typical cometary orbit as its crosses the Earth's orbit is well known, and this can be added to the Earth's speed—with a random factor to account for the unknown impact angle—to get the actual collision speed.

Once both the mass and speed are known, it's a simple matter to plug the numbers into the $\frac{1}{2}mv^2$ formula for KE, and take an appropriate fraction of this as the amount of destructive energy that will be transferred directly to the Earth. Since "destructive energy" is most commonly measured in megatons, the characters convert the result into those units and come up with the mind-bogglingly huge answer of 640,000 megatons. That's so much bigger than the largest nuclear weapons—which produce explosive yields of a few tens of megatons—that an even bigger unit of energy is needed. Using the fact that the giant Krakatoa volcanic eruption of 1883 was estimated at around 200 megatons, they conclude that their comet will have the destructive energy of around "3000 Krakatoas" [22].

Some of the equations we've met so far, such as that $\frac{1}{2}mv^2$ formula for KE, are really nothing more than definitions, while others—such as the equation Andy Weir uses for the motion of a pendulum—are just empirical rules of thumb. On the other hand, there are other equations that have a much stronger standing as fundamental "laws of physics". Including things like the conservation of energy and Newton's law of gravity, these may originally have been derived in laboratories on Earth, but they're incredibly powerful mathematical tools that can equally well be applied to phenomena anywhere in the universe.

It's largely because scientists have discovered these universal laws that we can say with reasonable confidence that we know "how space works". They're the one thing we can be reasonably certain of—because, as Chief Engineer Scott famously pointed out in a 1966 episode of *Star Trek*, you "can't change the laws of physics" [23].

1.4 The Laws of Physics

We've already met what may be the most basic physical law of all: the conservation of energy. This emerged from the work of a number of nineteenth century scientists, as SF author Charles Sheffield described in his short story "The Invariants of Nature" from 1993:

> Though different types of energy can be converted one to another, they decided that the total could never be changed. That was the principle of the conservation of energy [24].

The two forms of energy that loom largest in the context of space travel are kinetic energy and gravitational potential energy. We've said enough about KE (or $\frac{1}{2}mv^2$ for mathematicians) that its importance is clear enough. As for PE, that's a little harder to understand, particularly when we get out into space. But on or near the Earth's surface, it's simply proportional to the height above sea level. In other words, a stationary object high up on a hill has more energy than an identical object further down the hill.

Something that only a genuine physics nerd like Larry Niven would spot is that PE has serious consequences for that SF favourite, the teleportation booth. As he wrote in his 1974 short story "A Kind of Murder":

> Teleportation obeyed the laws of conservation of energy and conservation of momentum. Teleporting uphill took an energy input to match the gain in potential energy. A cargo would lose potential energy going downhill [25].

(We'll return to the other conservation law that Niven mentions, that of momentum, a little further on).

Once we get well away from the Earth's surface, the subject of PE becomes more complicated, for a couple of reasons. First, that high-school formula about mgh is only an approximation that works when the height h is small, and the formula for greater distances has to take into account the fact that the strength of Earth's gravity grows weaker the further away an object is. The second problem is that other bodies in the Solar System produce their own gravity has well, which has to be taken into account if it's significant enough (that's what Jules Verne did in the case of the Moon's gravity, in the calculation discussed earlier in this chapter).

When we get onto the scale of planetary orbits—or the orbits of other bodies that revolve around the Sun, such as comets or asteroids—the biggest contribution to PE comes from the Sun itself. The same basic principle applies here as to terrestrial PE; in other words, it's bigger the further away you are. For an object such as a comet that moves on a highly elliptical orbit, this has a consequence for the speed it moves at as it travels around its orbit. Conservation of energy means the comet has to trade KE for PE—so when then latter is large, at the most distant point of its orbit, its speed must be a minimum, and subsequently increase as the comet falls towards the Sun (Fig. 1.6).

It was this basic principle of physics that Larry Niven and Jerry Pournelle drew on when they estimated a comet's impact speed in *Lucifer's Hammer*, as discussed earlier.

A term that was used earlier in this chapter, without specifically defining it, was "escape velocity". It probably doesn't need definition, in fact—certainly not for SF readers or anyone interested in space travel—as it's virtually self-explanatory. In the case of the Earth, it's the speed a spacecraft needs to move at in order to break free from the Earth's gravity, such that it's no longer in orbit around it. If we rather unrealistically ignore the gravitational effect of everything else in the universe, then escape velocity is the speed at which an object has so much KE that, even having to trade this for PE the further and further away it gets, it can reach an infinite distance before its speed falls to zero.

It's also possible to talk about the escape velocity from the Sun's gravity, above which an object is no longer bound to an elliptical orbit within the Solar System. At the radius of the Earth's orbit, this is around 42 kilometres per second, and it's of more than academic concern to anyone interested in interstellar visitors (of either the animate or inanimate variety). As Arthur C. Clarke wrote in his non-fiction book *Report on Planet 3* (1972):

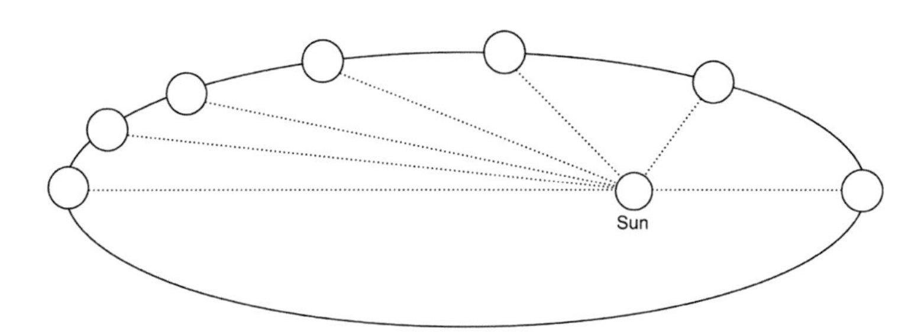

Fig. 1.6 Schematic illustration of an elliptical orbit around the Sun. The dotted lines are equally separated in time, so when they are bunched together at the furthest point, it means this is where the object is travelling slowest (Wikimedia Commons, CC-BY-3.0)

> In the Earth's neighbourhood, any object moving at more than [42 km/s] could only be a visitor to the Solar System, not a permanent resident. This is the limiting speed above which the Sun can no longer keep a body under its gravitational control. Anything moving faster than this, accordingly must have fallen into the Solar System from outside and would shoot out of it again after doing a tight turn round the Sun. [26]

When Clarke wrote those words, nothing moving that fast had ever been observed—but it has now, on more than one occasion. The first occurred in October 2017, when an asteroid-hunting team in Hawaii discovered an object that was, at the time, relatively close to Earth—yet moving at the unprecedented speed of some 50 km/s. This meant that its total energy—KE plus PE—was too high for it to be confined inside the Solar System. It was the Earth's first ever detected interstellar visitor (although almost certainly nothing more exciting than a rather hefty chunk of rock), that came in from the constellation of Lyra and then whizzed off again in the direction of Pegasus. In recognition of the Hawaiian team that discovered it, the object was given the Hawaiian name of 'Oumuamua, meaning "advance scout" [27].

The other famous conservation law, besides that of energy, is the conservation of momentum, as mentioned in the earlier quote from a Larry Niven story. Momentum is a less familiar term in everyday speech than energy, but in mathematical terms it simply means the mass of an object multiplied by its velocity, or mv. It's a useful quantity in many ways—not least because it now allows us, having mentioned Newton's first law of motion a few times, to describe the other two laws as well.

As mentioned earlier, the first law states that an object will always move in a straight line at constant speed unless there's a force acting on it. That's basically saying that, in the absence of an applied force, the momentum of an object remains constant—which is also the most obvious consequence of the conservation of momentum.

Newton's second law of motion states that if a force *is* applied to the object, then its momentum changes in such a way that the rate of change of momentum is numerically equal to the magnitude of the applied force. Finally, the third law says that, in Newton's own words, "To every action there is always opposed an equal reaction" [28].

The third law is another consequence of the conservation of momentum, which basically means that object A can't apply a force to object B without object B applying an equal and opposite force to object A. As dry and unexciting as this may sound, it's actually the essence of rocket science. The basic job of a rocket motor is to force a jet of hot gas out through a nozzle at the back—which would be a pointless exercise unless the jet of gas exerted an equal force on the rocket itself in the opposite direction (see Fig. 1.7).

Another way to understand how a rocket works is to use the conservation of momentum directly. When the rocket blasts out its propellant material, this material is carrying momentum with it, determined by its mass multiplied by the speed with which it is ejected. Since momentum has to be conserved, there is necessarily a "recoil" of the spaceship in the opposite direction to the ejected material, imparting an equal and opposite momentum to it. This will set it moving at a speed that is smaller than that of the exhaust by the same factor that the mass of the exhaust was smaller than the ship's mass.

Although conservation of momentum is one of the most basic principles of the universe, it's not something that many SF writers take the trouble to mention. One notable exception was Isaac Asimov, who discussed it explicitly in his 1952 short story "The Martian Way", in which spaceships use water as a propellant:

> It didn't take a ton of water to move a ton of ship. It was not mass equals mass, but mass times velocity equals mass times velocity. It didn't matter, in other words, whether you shot out a ton of water at a mile a second or a hundred pounds of water at 20 miles a second. You got the same velocity out of the ship [29].

Fig. 1.7 The upward force on a rocket as it lifts off is a direct consequence of Newton's third law of motion, because what the rocket is really doing is applying the same force *downwards* to its exhaust plume (Wikimedia Commons, CC-BY-4.0)

Here Asimov, like most people in the English-speaking world in the 1950s—and indeed most Americans even today—is using imperial measurement units, which can be confusing to users of the metric system. In the space of one sentence, he uses two different units to measure mass, "pounds" and "tons"—the latter being 2000 of the former. But there's a more subtle source of confusion here, too, because in strictly scientific terms, pounds and tons are units of weight, not mass.

There's a similar confusion in the metric system, too—where a person's weight, for example, is almost always measured in kilograms, even though that's technically a unit of mass. Before we confuse ourselves even further, let's take a step back and look at exactly what me means by "mass" and "weight"—and why, as soon as we venture off our own planet, the two are by no means the same thing.

Mass is a measure of the amount of substance in an object, and its main effect is to resist any change in the object's motion—in other words, to give it inertia. The higher its mass, the harder it is to start it moving, or to stop it again once it's in motion. This is encapsulated in Newton's second law of motion, which says that the force F needed to impart an acceleration a to an object is proportional to its mass m, i.e. $F = ma$.

The mass of an object is the same regardless of where it is in the universe—whether on Earth, or on a planet that's much larger or smaller, or on a spaceship in interstellar space. As just mentioned, the correct unit for its measurement in the metric system is the kilogram, but in the imperial system it's not the pound—but a unit most people who aren't engineers have never heard of, called the slug.

So where does a pound come in? It's most commonly encountered as a unit of weight, but it's really a unit of force, because weight *is* a force. The weight of an object is the force an object experiences due to whatever gravity field is acting on it. In other words, it's the F in $F = ma$ when a is the local acceleration due to gravity. A weight of 1 pound is the force acting on a mass of 1 slug when it's subjected to an acceleration of 1 foot per second per second. In the metric system, the unit of weight—or force—is the Newton, which is the force acting on a mass of 1 kilogram when it's subjected to an acceleration of 1 metre per second per second.

This may all sound unnecessarily complicated, and perhaps reminiscent of the more tedious kind of high-school physics lessons, but the difference between weight and mass is something we really have to get to grips with if we're going to understand how the laws of physics work in space. It's a point that Arthur C. Clarke made in one of his best-known novels, *2001: A Space Odyssey* from 1968:

> On the Moon the human body had to learn … for the first time, to distinguish between mass and weight. A man who weighed 180 pounds on Earth might be delighted to discover that he weighed only 30 pounds on the Moon. As long as he moved in a straight line at a uniform speed he felt a wonderful sense of buoyancy. But as soon as he attempted to change course, to turn corners, or to stop suddenly—*then* he would find that his full 180 pounds of mass, or inertia, was still there [30].

(There Clarke goes again, saying "a man" when he means "a person"—but at least he gets his physics spot on).

If you tried to lift a motorcycle off the ground—which isn't recommended—you'd be battling against its weight. On the other hand, leaving it on the ground and just pushing it along you'll still feel resistance—and properly speaking this resistance is due to its mass, or inertia, not its weight. You'd feel exactly the same resistance if you tried to push the bike inside a spacecraft where everything was weightless. That's an analogy Andy Weir draws in *Project Hail Mary*, when his protagonist is trying to move his unconscious alien companion:

> Rocky is heavy. Much heavier than I thought he would be. If there were gravity, I probably wouldn't be able to lift him at all. As it is, he has a lot of inertia. It takes a lot of *oomph* to pull him along. It's like pushing a motorcycle in neutral [31].

Some of the consequences of changing weight—and unchanging inertia—in different gravitational fields are counterintuitive. It's puzzling, at first sight, why astronauts on the Moon seem to move in slow motion even though they weigh so much less than on Earth. The situation is explained by a future lunar resident in Isaac Asimov's 1972 novel *The Gods Themselves*:

> When Earthmen … think of moving around the Moon, they think of the surface and of spacesuits. That's often slow, of course. The mass, with the spacesuit added, is huge, which means high inertia and a small gravity to overcome it…. Here, underground, without spacesuits, we can move as quickly as on Earth [32].

Now that we've come to terms with the difference between weight and mass, another question arises: how do you measure the mass of something? On Earth it's simply a question of weighing it—but when you're in a zero-gravity environment it's not that easy. Things no longer have weight, but they still have inertia, so you have to go by that. In his novel *The Apollo Murders*, Chris

Hadfield describes how an astronaut returning from the Moon estimates the mass of a collection of lunar rocks:

> He … swung the first box left and right, estimating the mass by how much it resisted his motion and twisted his upper body. "My guess is the furthest forward-stowed box weighs 50 or 60 pounds." Not exact, but the best he could do [33].

The best an astronaut could do, perhaps—but the science-teacher protagonist of Andy Weir's *Project Hail Mary* does a lot better, when faced with the problem of a metal ball of unknown mass:

> I put the half-depleted sipper into one of the buckets and the metal ball into the other. I set the whole thing spinning in the air. The two masses clearly aren't equal. The lopsided rotation of the two connected containers shows the water side is much heavier…. I pluck it out of the air and take a sip of water. I start it spinning again. Still off-centre but not as bad. I take more sips, do more spins, take more sips, and so on until my little device rotates perfectly around the marked centre point. That means the mass of the water is equal to the mass of the ball [34].

Weir's reference to a spinning object brings us neatly on to another important conserved quantity in physics, which goes by the ungainly name of "angular momentum". This is the rotational equivalent of ordinary linear momentum, and it's essentially a measure of the amount of spin that a system has. Though not particularly well known outside physics, angular momentum has played just as important a part in its history as the other conserved quantities, as Charles Sheffield points out in his story "The Invariants of Nature":

> In 1931 Pauli deduced the existence of a new particle, the neutrino, just because the principles of conservation of energy and angular momentum required it to exist [24].

Angular momentum also plays a key role in the theory of orbital motion; in fact the elliptical shape of most orbits comes from the need to simultaneously conserve energy and angular momentum around the orbit. The conservation of angular momentum also means that, once a spacecraft starts to spin, it won't stop until a suitable retarding force is applied. So, contrary to what many people may assume, the spin that's often applied to fictional space stations in order to create an artificial sense of gravity doesn't need a constant input of energy to maintain it. On the contrary, once it's started, you need to expend energy to stop it.

On the other hand, if only part of a space vehicle is spinning—such as the central carousel of the spaceship *Discovery* in Arthur C. Clarke's *2001: A Space Odyssey*, it's possible for this spin to be transferred to another part of the structure if it's physically coupled to it. That's what happens after *Discovery* is abandoned, as Clarke describes in the sequel, *2010: Odyssey Two*:

> Years ago, friction had braked the spin of *Discovery*'s carousel, thus transferring its angular momentum to the rest of the structure. Now, like a drum-majorette's baton at the height of its trajectory, the abandoned ship was slowly tumbling along its orbit [35].

From the way physicists use mathematical equations through to the fundamental conservation laws, this chapter has given us a glimpse of how physics works in outer space—and how this has been put to good effect by some of the more thoughtful SF writers. But there are still plenty of questions we haven't answered yet, such as just what gravity is, why it's so important to the way objects move through space—and how a rotating space station can give the illusion of gravity. We'll look at all these questions and more in the next chapter.

References

1. Artistic License–Space, *TV Tropes*, https://tvtropes.org/pmwiki/pmwiki.php/Main/ArtisticLicenseSpace
2. L. Niven, At the Bottom of a Hole, in *Tales of Known Space*, (Ballantine Books, New York, 1975), p. 99
3. B. Clegg, *Ten Billion Tomorrows* (St. Martin's Press, New York, 2015)
4. S. Webb, *All the Wonder that Would be* (Springer, Switzerland, 2017)
5. A.C. Clarke, *Odyssey two*, vol 1982 (Granada Publishing, London, 2010), p. 24
6. Scientific Inaccuracies. *The Martian Fandom*, https://the-martian.fandom.com/wiki/Scientific_Inaccuracies#The_Dust_Storm
7. A. May, *The Science behind Jules Verne's Moon Novels* (Post-Fortean Books, 2018), pp. 26–27
8. I. Asimov, *The Martian Way* (Panther Books, London, 1973), pp. 37–38
9. A. Weir, *Project Hail Mary* (Kindle Edition), loc. 1260
10. A. May, *The Science behind Jules Verne's Moon Novels* (Post-Fortean Books, 2018), pp. 79–80
11. C. Hadfield, *The Apollo Murders* (Kindle Edition), loc. 767

12. Archimedes' Principle, *Wikipedia*, https://en.wikipedia.org/wiki/Archimedes%27_principle
13. A. May, *The Science behind Jules Verne's Moon Novels* (Post-Fortean Books, 2018), p. 74
14. A.C. Clarke, A Meeting with Medusa, in *The Sentinel*, (Berkley Books, New York, 1986), p. 223
15. L. Niven, Becalmed in Hell, in *Tales of Known Space*, (Ballantine Books, New York, 1975), pp. 7–8
16. A. May, *The Science Behind Jules Verne's Moon Novels* (Post-Fortean Books, 2018), pp. 29–32
17. A. Weir, *Project Hail Mary* (Kindle Edition), loc. 578
18. A.C. Clarke, *The Sands of Mars* (Pan Books, London, 1964), p. 38
19. F. Hoyle, *The Black Cloud*, vol 46 (Penguin Books, London, 1960), pp. 26–27
20. A.C. Clarke, *Hammer of God* (Orbit Books, London, 1994), p. 184
21. A. May, *Cosmic Impact* (Icon Books, London, 2019), p. 126
22. L. Niven, J. Pournelle, *Lucifer's Hammer* (Futura Books, London, 1978), pp. 94–96
23. The Naked Time, *Star Trek* episode, first aired 29 September 1966
24. C. Sheffield, The Invariants of Nature, in *Analog Science Fiction and Fact*, (April 1993), pp. 21–22
25. L. Niven, A Kind of Murder, in *A Hole in Space*, (Orbit Books, London, 1984), p. 56
26. A.C. Clarke, *Report on Planet 3* (Corgi Books, London, 1973), p. 36
27. A. May, *Cosmic Impact* (Icon Books, London, 2019), p. 89
28. I. Newton, *Mathematical Principles of Natural Philosophy* (Adee, New York, 1846), p. 83
29. I. Asimov, *The Martian Way* (Panther Books, London, 1973), p. 45
30. A.C. Clarke, *A space odyssey*, vol 1968 (Arrow Books, London, 2001), p. 69
31. A. Weir, *Project Hail Mary* (Kindle Edition), loc. 4043
32. I. Asimov, *The Gods Themselves* (Panther Books, London, 1973), p. 174
33. C. Hadfield, *The Apollo Murders* (Kindle Edition), loc. 5680
34. A. Weir, *Project Hail Mary* (Kindle Edition), loc. 3406
35. A.C. Clarke, *Odyssey two*, vol 1982 (Granada Publishing, London, 2010), p. 69

2
Gravity

Almost everything that happens in space is affected by gravity. Its most obvious effect is to hold people and objects to the surface of a planet, but it's also the force that holds planets in orbit around a star, and satellites in orbit around the Earth. Even when travelling between planets, a spacecraft's trajectory is primarily governed by gravity. This chapter explains the basic physics of gravity, including why astronauts in orbit may appear to have escaped its influence even when they haven't. It also covers some of the practical consequences of gravity that are most frequently encountered in SF stories—from the stronger gravity of large planets and the artificial gravity of rotating space stations to Lagrange points, tidal forces and even that science-fictional favourite, the black hole.

2.1 Gravity on Other Worlds

For many people, gravity is simply the force that holds us to the surface of the Earth. In reality, however, gravity is much more than that. It's a force that pervades the entire universe, acting between any pair of objects in proportion to their masses and inversely proportional to the square of the distance between them. The reason we don't always think of it that way is that it's really a very weak force, so we only notice it if the other object—the one that's exerting its gravitational pull on us—has a very large mass. In practice, this only applies to the Earth, or to other astronomical bodies.

People whose minds work in a mathematical way—which, as we saw in the previous chapter, applies to many scientists and engineers—may want to do a

A. May, *How Space Physics Really Works*, Science and Fiction,
https://doi.org/10.1007/978-3-031-33950-9_2

little mental algebra at this point (everyone else can skip to the next paragraph). The volume of a spherical body is proportional to the cube of its radius r, so—if it's a constant density throughout—the same will be true of its mass. But we also said that the strength of gravity is inversely proportional to the square of distance, so putting those two things together means the gravitational pull on the surface of the sphere is actually proportional to $r^3/r^2 = r$.

In other words, for astronomical bodies of equal density, surface gravity is proportional to radius. That's only a rough guide, of course, because it's likely that a smaller body, having less force compressing it, also has a lower density. For example, the Moon's radius is just over a quarter of the Earth's, but its surface gravity—as we saw in the previous chapter—is only a sixth of Earth's. That's because it's made of lower density rock.

The principal effect of gravity is to produce an acceleration, which is a concept we used in the previous chapter without precisely defining it. Really it just means an increase in the speed of an object over a period of time—like the "zero to 100 km/h in X seconds" favoured by car manufacturers. Scientists usually measure acceleration in the rather awkward-sounding units of "metres per second per second", sometimes abbreviated as "metres per second squared" or m/s^2. A more intuitive unit, however, is g—the acceleration due to gravity at the Earth's surface, which is approximately 9.8 m/s^2. The Moon's gravity is around 1.6 m/s^2, or 0.16 g.

As we saw in the previous chapter, the one physics equation that everyone knows—even if they don't realize they know it—is distance equals speed multiplied by time. The corresponding equation for an accelerating object is a little harder, but if it starts from speed zero and accelerates at a constant rate a, then its *average* speed over time t is $\frac{1}{2}at$. We can then substitute this speed into the "distance = speed × time" formula.

That's what Ryland Grace, the protagonist of Andy Weir's 2021 novel *Project Hail Mary*, does when he wants to estimate the strength of gravity in an unfamiliar environment. It's a problem that might baffle many, but not Grace—who happens to by a high-school physics teacher. He carefully measures the time an objects takes to fall through a measured distance of 91 cm, and goes on from there:

> 0.348 seconds. Distance equals one-half acceleration times time squared. So acceleration equals two times distance over time squared.... I run the numbers and come up with an answer I don't like. The gravity in this room is too high. It's 15 metres per second per second when it should be 9.8. [1]

On Earth, we always feel gravity pulling us downwards, because that's where the largest amount of mass is. We don't feel mountains pulling us sideways, because their mass is negligible compared to the planet as a whole. But what if there was just the mountain, without the planet? In essence, that's what an asteroid or comet is. Take comet 67P, for example, the destination of the Rosetta mission by the European Space Agency (ESA) in 2014. In terms of its physical size and mass, 67P is roughly comparable to Mount Olympus in Greece or Mount Fuji in Japan [2].

There's no doubt that comet 67P has a gravity field of its own, albeit a weak one, as Rosetta's companion probe Philae (see Fig. 2.1) discovered when it landed there—after bouncing a couple of times. As ESA described the landing:

> The first touchdown was inside the predicted landing ellipse ... but then the lander lifted from the surface again—for 1 hour 50 minutes. During that time, it travelled about 1 kilometre at a speed of 38 centimetres per second. It then made a smaller second hop, travelling at about 3 cm/s, and landing in its final resting place 7 minute later [3].

Three decades earlier, the gravitational field of another comet—Halley, which at 11 km diameter is 3 or 4 times larger than 67P—was a topic of discussion in Arthur C. Clarke's novel *2061: Odyssey Three* (1987):

Fig. 2.1 Artist's impression of ESA's Philae probe, which showed that even a small comet has enough gravity for a landing (DLR German Aerospace Centre, CC-BY-3.0)

> Halley's gravity? Forget it—less than one centimetre per second squared—just about a thousandth of Earth's. You'll be able to detect it if you wait long enough, but that's all. Takes 15 seconds for something to fall a metre [4].

As so often, Clarke anticipated the practicalities of spaceflight long before they became a reality. In an even earlier story, "Summertime on Icarus" from 1960, he discusses the barely perceptible gravitational pull of an asteroid that's just 1.6 km across. As the title suggests, this is Icarus, a member of the "Apollo" group of asteroids that cross inside the Earth's orbit:

> The gravity of Icarus was 10,000 times weaker than Earth's; Sherrard and his space-pod weighed less than an ounce here, and once he had set himself in motion he floated forward with an effortless, dreamlike ease [5].

In the previous chapter, we encountered the concept of "escape velocity"—the speed needed to escape the gravitational influence of an astronomical body altogether. In the case of the Earth, we need powerful rockets to reach this velocity, but for an asteroid-sized object it may be more of a problem to keep your speed low enough that you *don't* inadvertently escape. This was another topic that intrigued Clarke, who referred to it in his non-fiction book *1984 Spring* in the context of one of the moons of Mars, Phobos, which has a diameter of around 23 km:

> Many years ago I read a science fiction serial in which one of the characters jumped off Phobos, the inner moon of Mars.... It can be said at once that Phobos (diameter about 15 miles) is too large a body to permit human beings to escape by muscle-power alone. However, there is no doubt that a man could jump off some of the smaller asteroids. The limiting diameter, for one made of ordinary rock, is about 4 miles [6].

In one of his earliest short stories, "Hide and Seek" from 1949, Clarke did actually have his protagonist stranded on Phobos:

> He stood in the centre of an irregular plain about 2 kilometres across, surrounded by low hills over which he could leap rather easily if he wished.... But unless he was careful, he might easily find himself at such a height that it would take hours to fall back to the surface [7].

That's distinctly reminiscent of Philae's experience on comet 67P, over 60 years later.

If tiny worlds have tiny gravity fields, then it's not surprising that planets that are more massive than Earth have stronger gravity. A good fictional example comes, once again, from Andy Weir's *Project Hail Mary.* His protagonist, Ryland Grace, encounters an alien hailing from a planet (which Grace dubs Erid) that orbits around the nearby star Epsilon Eridani. When the alien comes aboard Grace's spaceship, which at this point has an artificially induced Earth-normal gravity, this is the dialogue that ensues:

> Not much gravity. What is value?
>
> 9.8 metres per second per second.
>
> "Not much gravity," he repeats. "Erid gravity is 20.48."
>
> "That's a lot of gravity," I say. But that's to be expected. He'd told me all about Erid before, including its mass and diameter. I knew their surface gravity had to be roughly double Earth's [8].

Since we've already seen how surface gravity scales with the size and mass of a planet, we could probably have gone through the same reasoning process as Grace. So, of course, did Andy Weir before he wrote the book—and so would any good SF writer. Some of them go even further, doing pages of calculations "designing" their fictional worlds before writing a word of the story. One of the best-known examples of this is Hal Clement's classic novel *Mission of Gravity*, set on the lovingly crafted—if distinctly bizarre—fictional planet of Mesklin.

When the novel made its first appearance in 1953, it was in serialized form over the course of several issues of *Astounding Science Fiction* magazine. Perhaps the most remarkable feature of this serialization was a 13-page essay by Clement which accompanied the third instalment, in which he explained the physics behind the story. In essence, he "reverse-engineered" Mesklin to produce the strange gravitational effects he wanted, with the planet's gravity varying significantly over its surface due to its highly flattened, rapidly rotating form. The article even includes a hand-drawn diagram by Clement (see Fig. 2.2) showing the strength of Mesklin's gravity relative to that of Earth [9].

From a human point of view, the most tolerable region of Mesklin is near the equator, where the gravity is 3 times Earth normal. That's a huge strain, but might just be tolerable with sufficient training. At higher latitudes, however, the gravitational pull increases considerably, as the diagram shows—and

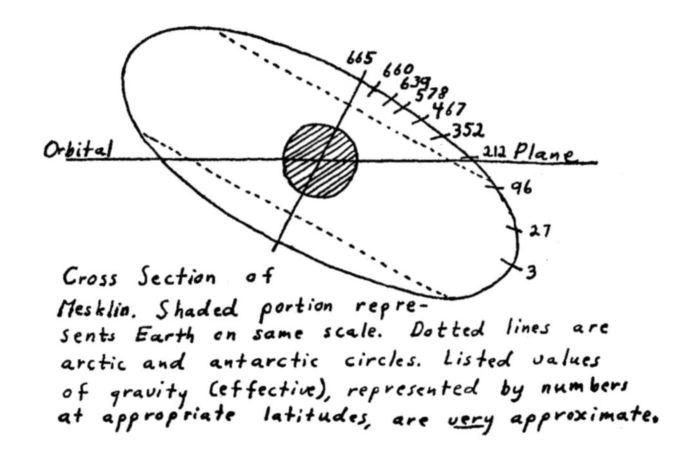

Fig. 2.2 Hal Clement's diagram showing how the surface gravity of his fictional planet Mesklin varies from 3 times that of Earth at the equator to 665 g at the poles (Internet Archive)

accordingly the indigenous lifeform has evolved to a centipede-like form that is very different from our own.

This contrasts with what might be the most obvious assumption, that—for example—aliens that were just like humans but three times larger would be ideally suited to a planet with three times the Earth's gravity. But physics doesn't work like that; physical properties don't always scale in direct linear proportion to each other.

The best-known example of this is the so-called "square-cube law", the effects of which we can see here on Earth in the relative strength of creatures of different sizes. The principle is explained, as well as anywhere, in a short story by Fletcher Pratt from 1952, conveniently titled "The Square-Cube Law":

> The square-cube law goes roughly something like this; as you increase the size, or mass, of an animal by the cube of its previous figure, its strength only goes up to the square. A man 30 feet high would be almost too weak to walk around.... On the other hand very small animals, like a mouse or a marmoset, are prodigiously strong for their size [10].

This idea may be difficult to grasp at first, but it gets easier if we think of actual squares and cubes—say a solid wooden cube, with six square faces, which could be either 1 or 2 cm on a side. The volume of the smaller cube is 1 cubic centimetre (1 cc), while that of the larger cube is 8 cc. Since they both have the same density, this means the larger cube weighs 8 times as much as the first. On the other hand, the side of the first cube has an area of 1 square centimetre, while the side of the second is just 4 times larger than this. In consequence, when they're placed side by side on the ground, the small cube has a weight of 1 unit pressing down on an area of 1 unit, while the larger one has a weight of 8 units pressing down on an area of 4 units—in other words, twice the amount of force per unit area.

The significance of this is that, as a general rule, the strength of an object—whether it's a wooden cube or a human body or anything else—generally scales with the square of its linear size, while its weight scales with the cube. So if its weight is suddenly increased by placing it in a stronger gravitational field, a small object will withstand the change more readily than a larger one. This means that, while creatures that evolve on higher-gravity planets will obviously need to be stronger, they will also benefit from being *smaller*—especially in the vertical dimension. Hence Clement's Mesklinites—and for that matter, Larry Niven's Jinxians, which he describes as "short and wide, with arms as thick as legs and legs as thick as pillars" [11]. We'll say a bit more about Jinx—a fictional moon orbiting a gas giant planet—later in this chapter, but the relevant point here is that its surface gravity is almost twice as strong as the Earth's.

2.2 Artificial Gravity

The other way in which gravity is likely to feature explicitly in SF stories, besides that of the weaker or stronger gravity of other worlds besides the Earth, is in the form of an artificially induced "pseudo-gravity" inside a spaceship or space station. Aside from purely fictional devices, there are two ways of doing this. We've actually met both of them already in this chapter, although the point wasn't emphasized at the time. First, there was the occasion in *Project Hail Mary* when Ryland Grace measured a pseudo-gravity of around 1.5 g—which, it turns out, was due to the fact that he was inside a rapidly accelerating spaceship. Then, later in the same novel, when the alien Rocky comes on board the same spaceship, it has an artificial gravity of exactly 1 g—due this time not to linear acceleration but to the fact that the spacecraft is spinning on its axis.

Before we examine these "artificial gravity" techniques in more detail, it's worth looking a little more closely at just how we experience gravity in the first place. Several times in this book we've described gravity as a "force", and while that's strictly true, it's not quite like any other force we might experience. Picture yourself in a jet aircraft as it's taking off. The aircraft is subjected to a horizontal force from its jet engines, acting in a forward direction. In accordance with Newton's second law of motion, which we met in the previous chapter, this force results in an acceleration of the aircraft and everything inside it, including yourself—and this acceleration likewise acts in the forward direction. But your *subjective* experience is in the opposite direction; you feel that you're being pushed backwards into your seat.

The force of gravity doesn't work in quite the same way. Suppose you're just sitting in a chair (which you probably are, as you read this) with the force of gravity acting downwards through the seat. In this case you're not accelerating at all, because the ground prevents you from doing that. Even so, you still have that subjective feeling of being pushed—in this case downwards into the chair, in much the same way that your were pushed back into your seat in the aircraft. But there's a distinct difference here—this time the push is in the *same* direction as the applied force, whereas in the aircraft it was in the opposite direction.

Let's just do one more thought experiment before trying to disentangle what's going on. This one is a little harder, because it's outside the experience of most people, but try to imagine what it would be like to fall freely under gravity with no ground beneath you to stop the fall. You may have felt something similar on a roller-coaster ride, or when a car goes too fast over a humpback bridge. In this case, you do undergo an acceleration—in the downward direction, the same as the force of gravity—but strangely you don't *feel* either the force or the acceleration. You just feel like you're floating, without the usual sensation of weight at all.

So there's no doubt we perceive gravity in a different way from any other force. The reason is that it acts equally on all material objects, imparting exactly the same acceleration to everything. So if gravity is unimpeded by any other force, as when we're falling freely, we're blissfully unaware of it. All the atoms of our body respond in exactly the same way, so we get no internal clues that any force is acting at all. When we talk about "feeling the pull of gravity", what we actually feel is the planet's surface pushing up against us and resisting the force of gravity.

When we see astronauts apparently "weightless" in the International Space Station (ISS), which orbits just 400 km above the Earth, it's not because gravity isn't acting on them. Its pull is almost as strong there as it is on the surface. And their weightlessness isn't literally because they're "falling", since they always remain at the same distance from the Earth. But orbiting objects are moving freely in the Earth's gravitational field in much the same way that falling objects are, so the effect is essentially the same. It's a point that Isaac Asimov made in his 1967 short story "The Billiard Ball":

> He was thinking of the kind of zero gravity one gets in a spaceship in free fall, when people float in mid-air. However, in a spaceship, zero gravity is not the result of the absence of gravitation, but merely the result of two objects, a ship and a man within the ship, falling at the same rate, responding to gravity in precisely the same way, so that each is motionless with respect to the other [12].

The weightless state of astronauts in orbit is often called "zero-g", not because gravity is absent, but because a person has no sensation of it. Conversely, it's possible to produce a gravity-like sensation in the absence of gravity—and this brings us back to the subject of artificial gravity. The most obvious approach to this is to exploit an effect we mentioned earlier—the way that an accelerating vehicle pushes us back into our seats rather like gravity pushing us down into the ground. It was this principle that caused Ryland Grace to feel a pull of 1.5 g when his spaceship was accelerating at that rate.

In the real world—as we'll see in the chapter on rocket science—it will be a long time before a spacecraft is able to maintain such a high acceleration for more than a few minutes. On the other hand, the general principle may well be possible at much smaller accelerations by, say, 2061. This was the assumption that Arthur C. Clarke made, anyway, in his novel *2061: Odyssey Three*:

> *Universe* was the first spaceship ever built to cruise under continuous acceleration ... she could manage a tenth of a g—not much, but enough to keep loose objects from drifting around [13].

The other approach to artificial gravity is to make use of the outward "centrifugal force" that objects feel inside a spinning structure. From a physics point of view, this can be seen as a consequence of Newton's laws of motion and the principle of inertia, but to be honest it's so familiar it really doesn't need a technical explanation. It's the principle behind a device as commonplace as a spin dryer, where rotation is used to force water outwards from freshly washed clothing. On a larger scale, rapidly rotating centrifuges are used to familiarize astronauts with the effects of high g-forces.

Centrifugally induced artificial gravity is most commonly encountered in SF in the form of rotating space stations. Perhaps the most famous of these is the one described in Arthur C. Clarke's 1968 novel *2001: A Space Odyssey*, and featured memorably in the movie based on it:

> Space Station 1 revolved once a minute, and the centrifugal force generated by this slow spin produced an artificial gravity equivalent to the Moon's [14].

The idea of a rotating, wheel-like space station wasn't originated by Clarke; it had been proposed as far back as 1952 in non-fiction article by Wernher von Braun—the German engineer who went on to design NASA's Saturn V rocket (Fig. 2.3).

Rotation has a huge advantage over constant acceleration as a source of artificial gravity, in that the latter requires a continuous input of energy to maintain it, whereas rotation doesn't. As we saw in the previous chapter, the fundamental principle known as the conservation of angular momentum means that once an object is set spinning, it will continue to do so at the same rate forever, or until a retarding force is applied to it.

An extreme example of this is seen in another of Arthur C. Clarke's novels, *Rendezvous with Rama* from 1973. "Rama" is the name given to a huge artificial cylinder that enters the Solar System, having presumably been constructed by an alien race in the distant past. It's virtually a self-contained world in its own right, which has been spinning on its axis for countless millennia. This creates an artificial gravity field that causes objects to fall *outwards* to the inside surface of the cylinder much as they would fall *downwards* to the surface of the Earth. Anyone in its interior is thus presented with a distinctly strange perspective, as the novel's protagonist discovers after he enters through the hub of one of the cylinder's end plates:

> He was not at the lowest point of this strange, inside-out world, but the highest. From here, all directions were *down*, not up. If he moved away from this central axis, toward the curving wall which he must no longer think of as a wall, gravity

Fig. 2.3 Wernher von Braun's original concept for a rotating space station (NASA image)

> would steadily increase. When he reached the inside surface of the cylinder, he could stand upright on it at any point, feet towards the stars and head towards the centre of the spinning drum. The concept was familiar enough; since the earliest dawn of space flight, centrifugal force had been used to simulate gravity. It was only the scale of this application that was so overwhelming, so shocking [15].

Although someone standing on the cylinder's inside surface might feel a very similar sensation to a person on the Earth's surface, there are subtle differences between this kind of artificial gravity and the real thing. These differences are most noticeable in the motion of falling objects, as Clarke's explorers find when they see a waterfall inside Rama:

> Descending from some hidden source in the clouds 3 or 4 kilometres away was a waterfall, and for long minutes they stared at it silently, almost unable to believe their eyes. Logic told them that on this spinning world no falling object could move in a straight line, but there was something horribly unnatural about a curving waterfall that curved sideways, to end many kilometres away from the point directly below its source [16].

The problem here is something physicists refer to as the Coriolis effect. Isaac Asimov gives a good description of this in his short story "For the Birds", from 1980. This is set on a rotating space station called Five, which is visited by an Earth resident named Modine. He gets a vivid demonstration of the Coriolis effect when he's persuaded to throw a ball into the air and catch it on the way down. This sounds simple enough, but as soon as the ball is airborne: "it curved parabolically, and Modine found himself drifting forwards in order to catch it, then running. It fell out of reach." The situation is explained to him by one of the station's inhabitants as follows:

> The difficulty is that what we call the Coriolis force is involved. Here at the inner surface of Five, we're moving quite rapidly in a great circle about the axis. If you throw a ball upwards it moves nearer the axis where things make a smaller circle and move more slowly. However, the ball retains the speed it had down here, so it moves ahead and you couldn't catch it. If you had wanted to catch it, you would have had to throw it up and back so that it would loop and return to you like a boomerang [17].

In the real world, of course, artificial gravity for space travellers is a luxury that still lies in the future. With current technology such as the ISS, it's simpler just to train astronauts to operate in a "weightless" environment. This does, however, have the downside that the general public often misunderstands the cause of this weightlessness. As our favourite high-school teacher, Ryland Grace from Andy Weir's *Project Hail Mary*, puts it:

> This is one of those things I frequently have to explain to my students. Gravity doesn't just "go away" when you're in orbit. In fact, the gravity you experience in orbit is pretty much the same as you'd experience on the ground. The weightlessness that astronauts experience while in orbit comes from constantly falling. But the curvature of the Earth makes the ground go away at the same rate you fall. So you just fall forever [18].

For an astronaut ascending to orbit atop a launch rocket, the feeling of weightlessness comes on suddenly as soon as the rocket's engine cuts off, as former astronaut Chris Hadfield describes in his 2021 novel *The Apollo Murders*:

> The third stage shut itself down exactly on time, and the smooth, easy push of its single engine was instantly gone. For the first time in their lives, the three men were weightless [19].

Hadfield was speaking from experience, as he flew in space himself on three occasions, the latter two taking him to the ISS (see Fig. 2.4).

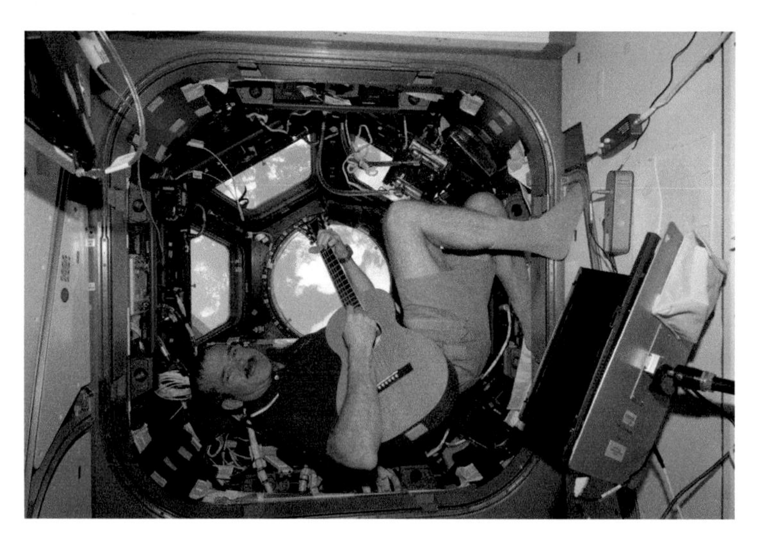

Fig. 2.4 Chris Hadfield, author of *The Apollo Murders*, spent a total of 166 days in space, so he knows exactly what weightlessness feels like (NASA image)

2.3 Orbits

Viewed on an astronomical scale, gravity has many other important effects besides that of holding objects to the surface of a planet. In fact, it's the driving force behind the whole field of celestial mechanics. Gravitational interactions between the Sun and planets, for example, are almost exclusively responsible for the entire structure of the Solar System. The basic concept at work here is that of an orbit, which is so crucial to our examination of "how space physics really works" that the whole of the next chapter will be devoted to it. But it's worth giving a brief introduction to the subject here, before we move on to a few other effects of gravity that crop up in SF stories.

In essence, an orbit is the trajectory of any object that is moving primarily under the gravitational influence of one or more other objects. Strictly speaking, this means that even when you throw a ball—or fire a bullet from a gun—it's moving in orbit around the centre of the Earth. But the orbit has so little energy that it hits the ground long before it completes a revolution of the planet. On the other hand, what would happen if the ball, or bullet, had a lot more energy? This was a question that Isaac Newton asked himself in his famous book *Principia Mathematica*, from 1687. Specifically, he considered an idealized "thought experiment" in which a cannonball is fired with increasing horizontal speed from the top of a high mountain (see Fig. 2.5).

This experiment would never work in the real world, because air resistance—even at the height of the world's tallest mountain—would slow the cannonball down. But if we imagine the cannonball being given its horizontal velocity at an altitude of several hundred kilometres—having been launched there by a rocket, for example—then Newton's argument is perfectly valid. At this point, the only force acting on the cannonball is gravity, which is constantly pulling it downwards towards the centre of the Earth. So, when the horizontal speed is low, the cannonball behaves exactly as we would expect, falling more or less straight back to the ground. That's the case labelled D in Newton's diagram.

As the horizontal speed is increased, however, the cannonball travels further and further before gravity manages to pull it back down to Earth. But when we remember that the Earth is a sphere, we see that "travelling further and further" actually equates to "curving further and further around the planet"—as shown in cases E, F and G. Eventually, when the horizontal speed is high enough, the cannonball will travel all the way round the Earth without ever hitting it. And this, of course, is what we usually understand by the word "orbit".

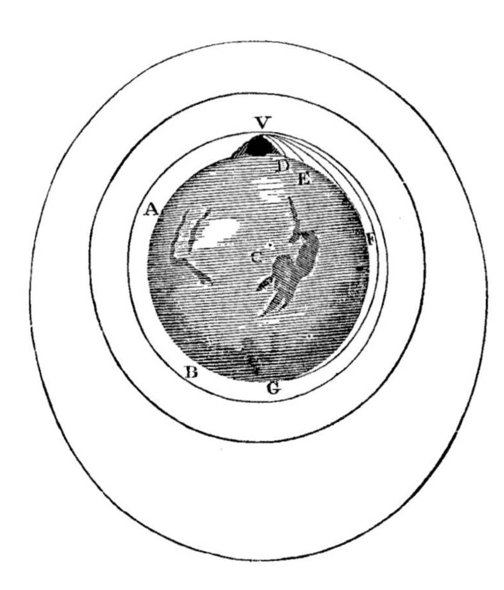

Fig. 2.5 An illustration from an 1846 edition of Isaac Newton's *Principia*, showing how a cannonball launched from a high mountain V travels increasing distances to points D, E, F and G as its speed is increased. At even higher speeds, the cannonball goes into orbit around the Earth (public domain image)

Perhaps the most surprising point here is that we're only putting energy into the cannonball—or whatever else we're launching into orbit—right at the start of the scenario, to give it the necessary horizontal speed (and, in the real world, to lift it to a high enough altitude that we can ignore the effects of air resistance). After this point, it doesn't require any effort at all to keep something in orbit. In fact, in the absence of any frictional effects, the main thing that does require effort is bringing it back down to Earth again—using braking rockets, for example.

In spite of the fact that Isaac Newton worked all this out almost 350 years ago, it remains a mystery to many non-specialists why objects don't have a natural tendency to fall out of orbit. This deeply ingrained misconception allows the hero of Arthur C. Clarke's 1953 story "Jupiter V" to work a clever bluff. He not only threatens to drop the space-suited villain from a Jupiter-orbiting satellite, but then proceeds to carry the threat out. Yet the villain was never in any real danger, as the protagonist eventually concedes:

> When I said that a body would take 95 minutes to fall from here to Jupiter, I omitted—not, I must confess, accidentally—a rather important phrase. I should have added "*a body at rest with respect to Jupiter.*" Your friend was sharing the orbital speed of this satellite—and he's still got it.... He's still moving in practi-

> cally the same orbit as before. The most he can do … is to drift about 100 kilometres inward. And in one revolution—12 hours—*he'll be right back where he started*, without us bothering to do anything at all [20].

This demonstrates another characteristic feature of orbits—that they're easily predictable, and you can always calculate where an orbiting object will be at any future moment. As Clarke says in another story, "The Other Side of the Sky" from 1957, "Nothing is ever really lost in space. Once you've calculated its orbit, you know where it is until the end of eternity." [21].

There's an amusing real-world anecdote along these lines that's related by Bergita and Urs Ganse in their *Spacefarer's Handbook*:

> While performing a spacewalk for repairs on the International Space Station in November 2008, the American astronaut Heidemarie Stefanyshyn-Piper forgot to clip a tool bag onto a holding rail on the station's surface. The bag slowly started to drift away, and the astronaut only realized this when it was out of reach. The bag continued to fly by itself on an orbit in proximity to the station and came close to it again 45 minutes later, but the astronaut was unable to climb the outside of the station fast enough to catch it.… The lost tool bag circled Earth on its own for a couple of months while slowly sailing away from the ISS. It was photographed from the ground by amateur astronomers and used as a test object for space debris tracking. It eventually got slowed down by the tenuous atmospheric friction, re-entered in August 2009 and burned up [22].

The only way a spacecraft can change from one orbit to another is by firing its rocket motors—and interestingly enough, there are some natural objects that do something very similar. In the previous chapter, we encountered the wayward comet that poses a threat to the Earth in the 1977 novel *Lucifer's Hammer* by Larry Niven and Jerry Pournelle. In common with many real-world comets (see Fig. 2.6), this has its own brand of "rocket motor" built into its basic constitution:

> The comet's made of rocks and dust, the dirt balled up with ices and frozen gases.… Pockets of these gases thaw and blast out to one side or the other. Like jet propulsion, and it changes the orbit [23].

One real-world example of cometary jet propulsion was observed in 1992, just in time for Arthur C. Clarke to slip a brief mention of it into his novel *The Hammer of God*, which was published the following year. Clarke describes how comet Swift-Tuttle, first discovered in 1862, "was then lost for more than

Fig. 2.6 When a comet heats up as it approaches the Sun, the gases expelled into its tail can alter the course of its orbit through a rocket-like effect (NASA image)

a century because … its orbit had been changed by jet reaction as it approached the Sun". It was subsequently rediscovered in 1992, he continues:

> And when its new path was computed, there was widespread alarm. It appeared that Swift-Tuttle had a high probability of hitting the Earth on 14 August 2126. Although this created a sensation at the time, the episode is now virtually forgotten. When the comet rounded the Sun in 1992, its solar-powered jet changed its orbit again—to a safe one [24].

2.4 The Three-Body Problem

Virtually the whole of gravitational physics is encapsulated in a single equation in Newton's *Principia*, which relates the force between two bodies to their masses and the distance between them. There are a few exceptional situations that call for a more sophisticated theory developed by Einstein, but—apart from a brief look at the topic of black holes at the end of this chapter—we're not going to encounter any of those situations in the space-travel scenarios considered in this book. So does this mean that Newton understood all the non-Einsteinian gravitational scenarios that have ever been studied since his time? No it doesn't—far from it.

It sounds like a contradiction, saying on the one hand that all the necessary physics is contained in Newton's equation, while on the other hand, Newton himself only understood a small fraction of this physics. Yet both statements are true. It's possible to have an equation that's so difficult to solve that only a tiny number of solutions—in simple, idealized situations—are actually known. This was the case with Newton's theory of gravity prior to the age of digital computers.

If you look at it closely, all the neat mathematics that's taught in high schools and universities about Newtonian gravity only ever deals with the problem of two bodies—for example, the Earth orbiting around the Sun, or a satellite orbiting around the Earth—while ignoring the gravity of every other object in the universe. That's because Newton's equation has a single, exact solution for the two-body problem—but not for any other configuration. As soon as a third body is added, you almost always need a computer to work out the answer on a case-by-case basis. As Ian Stewart wrote in his book *Does God Play Dice?*:

> In human affairs, two's company and three's a divorce. In the same way, in celestial mechanics the interaction of two bodies is well behaved, but that of three is fraught with disaster [25].

Before computers came onto the scene, the traditional approach to the three-body problem was to take the standard two-body solution, and then treat the third body as a relatively small perturbing influence on it. The first great triumph of this approach was the discovery of the planet Neptune in the nineteenth century. The basic principle behind this discovery was described by the science writer Willy Ley in, of all places, the August 1956 issue of *Galaxy Science Fiction* magazine. While Ley wasn't an SF author himself, he mingled closely with people who were, and contributed regular science-fact articles to *Galaxy* throughout the 1950s and 60 s. Here's what he wrote on the subject of the three-body problem:

> Consider the case of a single planet moving around a sun.... Now we add a second planet, which moves around the same sun, but in an orbit outside the orbit of the first planet. It will move at a slower rate and it also has to follow a much larger path. The result is that the inner planet will overtake the outer one at regular intervals. As the two come near each other, the effects of their own gravitational fields will enter into the game. They pull each other, and the inner

> planet moves a bit faster than it would if the other did not exist. By the same token, the outer planet is slowed down a bit. The closer they come to each other, the more strongly this mutual "perturbation" will show up.... In short, if a planet, at a certain point of its orbit, first speeds up and then slows down, it indicates a gravitational pull by a body in an orbit farther away from its sun. [26]

In the case of Neptune, the planet whose orbit it was perturbing was Uranus, the odd behaviour of which convinced several astronomers that there had to be another planet's gravity acting on it. As Ley says:

> In France, Urbain Leverrier presented his report on the calculation of the existence and position of a trans-Uranian planet to the French Academy on November 10, 1845. In England, John Couch Adams ... did the same work and forwarded his report to Sir George Airy around November 1st, 1845. There has been much unnecessary discussion on whether priority should be awarded to Adams or to Leverrier. The plain fact is that both men did the same work at the same time. Since astronomical circles were well acquainted with the "misbehaviour" of Uranus, it is surprising that more people were not attacking the same problem simultaneously [26].

In a science-fictional context, similar orbital perturbations feature in Fred Hoyle's novel *The Black Cloud*, from 1957. As described in the previous chapter, this deals with the discovery of a large interstellar gas cloud approaching the Solar System. One of the characters in the novel—who, like Hoyle himself, was a professional astronomer—used the method of perturbations to estimate the mass and location of the cloud. As the character says, this amounts to "much the same thing as the J. C. Adams—Leverrier determination of the position of Neptune." [27].

Another instance where the perturbing effect of a third body becomes important is when a large planet changes the orbit of an asteroid. This is something that happens from time to time in the real world—and also in at least one SF novel, *Shiva Descending* from 1979. This was a collaborative effort by two authors: Gregory Benford—who like Fred Hoyle, was a professional astronomer at the time—and William Rotsler. In the novel, Shiva is an Apollo asteroid, like the real-world Icarus mentioned earlier in this chapter. This means that it moves on a relatively eccentric orbit, with an outermost point near the orbit of Jupiter but the innermost close to the Earth's. It's a situation where the perturbing effect of Jupiter's gravity can cause a slow but steady change in the asteroid's orbit. This has disastrous consequences in the case of the fictional Shiva, which gradually shifts from a highly inclined orbit—which poses no threat at all to the Earth—to one that's on a collision course with our planet [28].

Calculations of orbital perturbations are really only for the most astrophysically dedicated SF writers, like Fred Hoyle and Gregory Benford. On the other hand, another facet of the gravitational three-body problem has become so well known—though perhaps not in quite those terms—that it's a common occurrence in SF today. This is the phenomenon of "Lagrange points": specific solutions of the three-body problem that are possible under certain restrictive circumstances, and which result in very stable orbits. The Lagrange solutions relate to situations where the first two bodies—which may be the Sun and the Earth, for example, or the Earth and the Moon—are very much more massive than the third body, which in the SF context is usually a spacecraft or space station.

The basic idea is explained by Arthur C. Clarke in his "science-fictional autobiography", *Astounding Days*, from 1989:

> One of the universe's greatest challenges to human mathematicians [is] the problem of three bodies. In general there is no exact solution, though there are approximation methods which will give the answer to any desired degree of accuracy.... One of the greatest mathematicians who ever lived, Lagrange, discovered that for one special case the problem had a solution—or rather, 5 solutions. If the mass of the third body is negligible compared with the other two ... then it can remain in equilibrium at (or near) 5 fixed points [29].

The 5 points in question are illustrated in Fig. 2.7. If an object is situated precisely at one of these "Lagrange points", then its position will remain forever fixed relative to the two larger masses. More realistically, if an object is

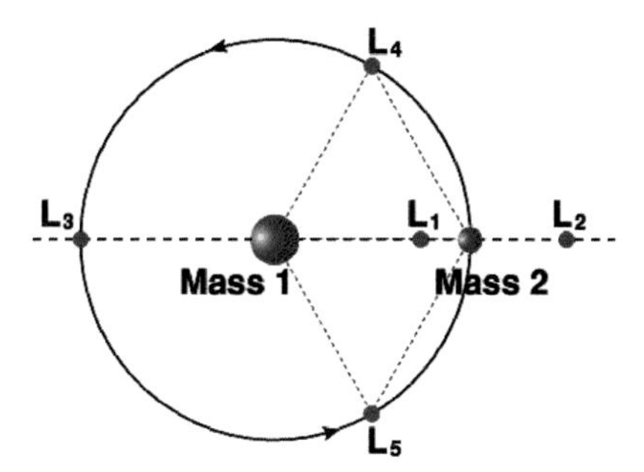

Fig. 2.7 A schematic illustration of the 5 Lagrange points in a system that is dominated gravitationally by two larger masses (NASA image)

close to a Lagrange point but slightly off, it will oscillate around that point without ever departing very far from it.

Of the 5 Lagrange points, the easiest to understand is L1. In the Earth-Moon system, for example, this is simply the neutral point on the straight line joining the two, where the pull of gravity from each of them has the same magnitude but opposite direction. This point was mentioned (though not by name) in the previous chapter, in the context of Jules Verne's calculation of Earth-Moon flight dynamics. The other 4 Lagrange points are less intuitive, but suffice to say the same kind of gravitational balance applies to them as well.

In the real world—and in the context of the Sun-Earth, as opposed to Earth-Moon, system—the L2 point has become prominent in recent years because several space observatories, most famously NASA's giant James Webb Space Telescope, are located there. This point has the special feature that it's always on the exactly opposite side of the Earth from the Sun, so that a telescope located there always sees the Sun and Earth is the same direction. This allows it to screen itself simultaneously from the two biggest sources of interference—whether heat, light or radio waves—with a single sunshield.

In science fiction, on the other hand, it's the L4 and L5 points that are most often encountered. As the diagram shows, these form the third point of an equilateral triangle with the other two bodies. As with the other Lagrange points, a body placed at either of these points will remain essentially fixed there relative to the two larger bodies. One of the first SF writers to exploit this effect, in the context of the Sun-Venus system, was George O. Smith in his (at one time) popular *Venus Equilateral* series. The setting here is a space station that lies in the same orbit as Venus, but located at the L4 point—so that it perpetually remains 60 degrees ahead of the planet as they revolve around the Sun. As Smith explains in the first story in the series, "QRM Interplanetary" from 1942:

> One of the classic solutions of the problem of the Three Moving Bodies, in which it is stated that three celestial objects at the corners of an equilateral triangle will so remain, rotating about their common centre of gravity [30].

While L4 travels 60 degrees ahead of the secondary mass, L5 is its mirror image, travelling 60 degrees behind the secondary. In the context of the Earth-Moon system, the L5 point attracted attention in the 1970s—not least in the pages of SF—as a potential location for a long-term space habitat. As *The Encyclopedia of Science Fiction* relates:

> The Princeton physicist Gerard K O'Neill, an important propagandist for space colonies, argued in *The High Frontier* (1977) that good sites for such colonies

> would be L4 and L5 of the Earth-Moon system, 60 degrees ahead of and behind the Moon in its orbit.... He particularly liked L5, and this region soon became something of an SF cliché as the site for fictional space cities consisting of clusters of self-supporting habitats [31].

O'Neill had a couple of reasons for favouring the L4/L5 points. For one thing, his space habitats had to be close enough to the Earth to be readily accessible, but far enough away—and in a sufficiently stable orbit—that there was no risk of them crashing into the planet. At the same time, the material for their construction should preferably come from somewhere with a lower gravity than the Earth, making it easier to transport into space—and the Moon was an ideal candidate. So he settled on those two Lagrange points, which are conveniently close to both the Earth and the Moon, while also being super-stable from a gravitational point of view.

Interestingly, given that O'Neill's book was written just a few years after Arthur C. Clarke's *Rendezvous with Rama*, his favoured design for a space habitat was a large cylinder that rotated on its axis to produce artificial gravity, not at all unlike Clarke's Rama. Now referred to as an "O'Neill cylinder" (see Fig. 2.8), a typical habitat would be 8 km in diameter and 32 km in length, giving a total internal surface area of just over 800 square kilometres. That's

Fig. 2.8 Artist's impression of the interior of an "O'Neill cylinder" space habitat (NASA image)

slightly larger than the island of Singapore, with its population of 5.7 million, so O'Neill's space habitats were nothing if not ambitious [32].

Needless to say, gravitational systems don't stop with three bodies, and it's possible to find configurations of four or more that allow stable solutions analogous to the Lagrange points. One particularly elegant solution, popularized by Larry Niven in his 1970 novel *Ringworld*, is the "Klemperer rosette" (although Niven managed to misspell Klemperer as "Kemplerer"). In Niven's story, this is the configuration that an advanced race of aliens known as "Pierson's Puppeteers" employ when they re-engineer their home system. In this instance the rosette consists of 5 bodies, although any number is possible in theory. Here is how Niven describes it in the novel:

> Take three or more equal masses. Set them at the points of an equilateral polygon and give them equal angular velocities about their centre of mass. There the figure has stable equilibrium [33].

2.5 Tidal Forces

There's one other gravitational effect that crops up often enough in SF to be worth discussing in some detail, and this relates to tides. That's a familiar enough word, even in a non-scientific context, when it's used to refer to the twice daily ocean tides on Earth. In fact, for hundreds of years this was the only meaning of the word, until Newton explained the underlying physics of the phenomenon in the seventeenth century. He showed that, as a few people before him had suspected, the tides are due to the effects of the Moon's gravity on the Earth. But what's at work here isn't simply the gravitational force exerted by the Moon on the Earth, but the difference in that force—technically known as the "gravitational gradient"—as it's felt by different parts of planet lying at different distances from the Moon.

There's no doubt that the Sun's gravity pulls on the Earth much more strongly than the Moon's—that's why the Earth orbits around the Sun, while the Moon orbits around the Earth. So why is it the Moon, rather than the Sun, that causes the tides? The fact is the Sun does have a tidal effect on the Earth, but it's much smaller than that of the Moon, because it's so much further away. We've seen that the force of gravity falls off as the inverse square of distance, but it turns out the gravitational gradient—the thing that causes the tides—falls off much more quickly as the inverse *cube* of distance. That's enough, despite the fact that the Sun has so much more mass than the Moon, for the Moon to win this particular battle and end up as the principal cause of Earthly tides.

The basic effect of a tide—or of a gravitational gradient, if it's easier to visualize in those terms—is to try and pull an extended object apart. That's because the gravitational force acting on the nearer side is greater than the force on the more distant side. In the case of the Moon's tidal effect on the Earth, this causes a bulge in the oceans (see Fig. 2.9)—but the stronger tides imagined by some SF authors in other situations can have much more destructive effects.

As the Moon example shows, the key to getting strong tidal effects doesn't lie in the mass of an object, but in its *distance*—specifically, how close it is relative to the size of the object you want it to affect. To be really engaging in an SF context, the latter object can't be something as large as a planet, but ideally needs to be a spaceship or even a human being. So we're really looking for a tidal force, or gravitational gradient, that's strong enough to tear a person apart.

That's never going to be the case with any object made of ordinary matter, such as a moon, planet or star. In such cases, the gravitational gradient reaches its maximum value on the surface of the object—and, as we know from living on the surface of the Earth, it's nowhere near strong enough to pull anything apart. What we need instead is some exotic form of matter than is far denser than the ordinary kind. Fortunately for SF writers, something of this nature really does exist in the universe, in the form of neutron stars. These have a mass comparable to that of the Sun, but all compressed down into a sphere that's only a few kilometres across.

Normal matter is made up of atoms, and these in turn are composed of three types of subatomic particle—electrons, protons and neutrons—with a lot of empty space between them. But if these particles are squashed together

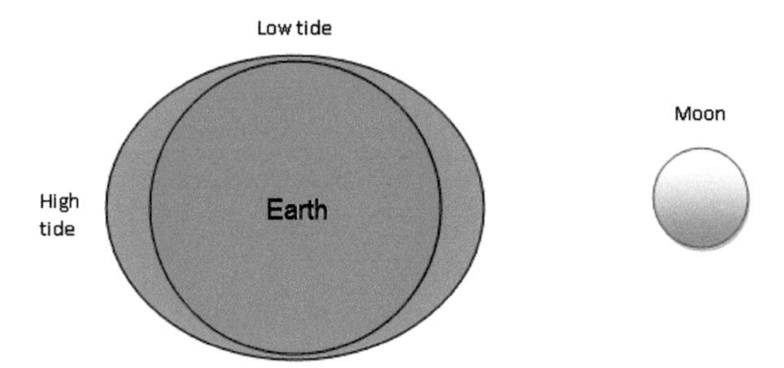

Fig. 2.9 A schematic view of the Moon's tidal effect on the Earth. Imagine we're looking down on the Earth, which rotates once per day. As it does so, the tidal bulge passes each point on Earth *twice* a day (Wikimedia Commons, CC-BY-SA-4.0)

forcefully enough, then all the protons and electrons combine with each other to form further neutrons, which can be pushed so tightly together that they're virtually touching. In the end, this highly compressed matter is all that's left—and it's the end state of a certain category of stars when they run out of nuclear fuel. At this point the star collapses down under its own gravity, and if it's massive enough to continue collapsing beyond the white dwarf stage (but not massive enough to collapse all the way down to a black hole, which we'll come to a little later), then a neutron star is the inevitable result. As Larry Niven wrote in his award-winning short story "Neutron Star", from 1966:

> Electrons would be forced against protons to make neutrons. In one blazing explosion most of the star would change from a compressed mass of degenerate matter to a closely packed lump of neutrons: neutronium, theoretically the densest matter possible in this universe [34].

The point about a neutron star, compared to any object made from ordinary matter, is that the combination of a large mass and a small size mean you can get extremely close to the centre of gravitational attraction—close enough for the tidal gradient to become a real hazard to space traffic. As Arthur C. Clarke put it, in a tongue-in-cheek story called "Neutron Tide" from 1970:

> In any reasonable gravitational field—even that of a white dwarf, which may run up to a million Earth g's—you just swing around the centre of attraction and head out into space again, without feeling a thing.... For a neutron star, however, this is no longer true. Near the centre of mass the gravitational gradient—that is, the rate at which the field changes with distance—is so enormous that even across the width of a small body like a spaceship there can be a difference of 100,000 g's [35].

Clarke's story is really just an excuse for him to conclude with a spoonerism-style punchline about a "star-mangled spanner". On the other hand, Larry Niven's "Neutron Star" treats the same basic scenario in a much more serious way. His protagonist, Beowulf Shaeffer, is almost pulled apart by tidal forces when his spaceship ventures too close to a neutron star called BVS-1. As Shaeffer explains later:

> Picture it. The ship's nose was just 7 miles from the centre of BVS-1. The tail was 300 feet farther out. Left to themselves, they'd have gone in completely different orbits. My head and feet tried to do the same thing when I got close enough [36].

We'll come back to the destructive potential of tidal forces a little later, when we get on to the subject of black holes, but for now we'll turn to another tidal effect that crops up now and again in SF. This is the phenomenon of "tidal locking", and its most obvious consequence for us here on Earth is the fact that the Moon always keeps the same side pointed towards us. As Chris Hadfield observes in *The Apollo Murders*: "Only 24 humans—all Apollo astronauts—have seen the other side." [37].

Just why the Moon should be "tidally locked" in this way isn't obvious at first glance, but we can start by observing that the Earth's tidal effects on the Moon are going to be larger than the Moon's effect on us. That's because the Earth has more mass than the Moon, while its distance away is exactly the same. The effect, over billions of years, has been to create bulges on either side of the Moon, by distorting the rock it's made of in the same way the Moon itself distorts the oceans on Earth. The differential pull of Earth's gravity on these rocky bulges then slowed down the Moon's rotation, to the point where it always keeps the same side facing us.

Tidal locking is a widespread astronomical phenomenon, found in many cases where a small body orbits relatively close to a larger one. In the Solar System, the two Martian moons Phobos and Deimos are both tidally locked to Mars, as are the moons Ganymede, Callisto, Io and Europa, among others, to Jupiter. There are also tidally locked moons orbiting around Saturn, Uranus and Neptune.

On the other hand, the planet Mercury isn't tidally locked to the Sun, even though most astronomers believed this to be the case prior to the 1960s. If it had been, it would have created the intriguing situation where one half of the planet was in perpetual sunlight and the other half in never-ending darkness. This raises a whole range of fascinating fictional scenarios, so it's not surprising that many early SF stories set on Mercury focus on this supposed "fact" about the planet.

One classic example is a science-fictional murder mystery by Isaac Asimov, "The Dying Night", dating from 1956. The key to the mystery's solution depends on Mercury being tidally locked—so, now that we know it isn't, the story no longer makes any sense. Asimov later admitted that he'd tried to fix the problem, but given up because he couldn't work out how to do it without changing the plot beyond recognition [38].

Another victim of the same misconception was Larry Niven's first published story, "The Coldest Place", from 1964. The title is a reference to the side of Mercury that, as far as Niven knew when he wrote the story, constantly faces away from the Sun. Unfortunately for the story's longevity, astronomical progress caught up with it very quickly. As Niven himself put it, "'The Coldest Place' was obsolete before it ever reached print" [39].

Even if Mercury didn't live up to expectations, the idea of tidally locked worlds is so full of potential that several authors have invented such worlds for themselves. One example, mentioned earlier in this chapter in the context of its abnormally high gravity, is Niven's fictional moon Jinx. This has a shape almost as bizarre as Hal Clement's Mesklin:

> Tidal force had earlier locked Jinx's rotation to Primary and forced the moon into an egg shape, a prolate spheroid.... This is why the ocean of Jinx rings its waist, beneath an atmosphere too compressed and too hot to breathe; whereas the points nearest to and furthest from Primary, the east and west ends, actually rise out of the atmosphere [40].

In yet another Niven story, "One Face" from 1965, the crew of a spaceship find themselves flung into a far-distant future where tidal effects have taken their toll on the Earth itself—which, accordingly, "has become a one-face world; it turns one side forever to the Sun." [41].

2.6 A Few Words about Black Holes

We said earlier that virtually the whole of gravitational physics can be understood using Newtonian theory. As far as real-world astrophysics goes, the most important exception to this is the existence of black holes. It's probably no coincidence that black holes also happen to be by far the most popular astrophysical phenomena found in science fiction. So what exactly is a black hole, why is it such an appealing concept for SF writers, and how does the real-world variety differ from its fictional counterpart?

To put it as simply as possible, a black hole is a point in space that has a finite mass but no spatial extent—in other words, it's a point of infinite density. Larry Niven was perfectly correct when we quoted him earlier as saying that a neutron star is "theoretically the densest matter possible in this universe"—because a black hole, with its infinite density, isn't made of matter at all, in the sense of atoms or subatomic particles such as neutrons. It's a phenomenon that's unique in itself: a pure point mass.

To properly understand the physics of black holes, we would need to go beyond Newton's law of gravity to Einstein's much more complicated theory of General Relativity, which deals with gravitational fields that are so strong they can warp the very fabric of space and time. Put like this, it's no surprise that black holes have such appeal for SF writers; they open up so many possibilities. Even Einstein himself, in collaboration with Nathan Rosen, speculated that black holes might act as bridges, or shortcuts, between one point in space-time and another.

For many SF authors, that's all they need to know, and black holes are frequently put to work as convenient "portals" allowing instantaneous travel to another galaxy, another period of time or even a parallel universe. But the physics of such portals is shaky at best—and even if it wasn't, black holes are never going to be a practical means of transport due to their very nature. The reason is obvious if we think back to the preceding discussion of tidal forces.

If you're going to travel *through* a black hole, you first need to get within zero distance of it—and long before this you'll be subjected to tidal stresses that make a neutron star look puny in comparison. The technical term for what happens to an object falling into a black hole is "spaghettification", which is more or less self-explanatory. The intense tidal forces will stretch the object into long, thin strands like spaghetti [42].

If black holes are never going to make a practical form of interstellar transport, they're still interesting for their purely gravitational effects—which, sadly, often receive an unrealistic portrayal in fiction. We'll look at the facts and fallacies of this in a moment, but first it's worth considering where black holes come from, and where they might be found—by real-world spacefarers just as much as science-fictional ones.

Historically, the idea of a black hole, much like that of a neutron star, originally arose in the context of stellar collapse—and it's still frequently presented in this way. We said that neutron stars are one possible result of the gravitational collapse that ensues after a star runs out of nuclear fuel. They're actually the middle case of three possibilities. At the low-mass end of the spectrum, a star like our own Sun will end its days as a white dwarf, where the inward collapse due to gravity is halted when all the atoms in the star are compressed into their densest possible form. With a star that's a little more massive, however, gravity is able to compress the atoms themselves to produce a neutron star. And the most massive stars of all don't even stop there; they continue collapsing all the way down to a single point, i.e. a black hole.

This is the main process by which new black holes can be created in the universe today. It results in a black hole with a mass comparable to that of large star—but this doesn't mean that black holes of other sizes can't exist. It's believed that, long ago in the very early history of the universe, large numbers of them could have arisen spontaneously. These "primordial black holes" could be almost any size, from the mass of a galaxy down to the mass of an asteroid. At the lower end of this scale, it's by no means impossible—though not particularly likely—that astronauts of the future might genuinely encounter a black hole on their travels. According to some estimates, there may be as many as one black hole for every 10 visible stars in the universe [43].

So what would happen if a spacecraft really did come close to a black hole? Fortunately, we can answer this question using the familiar gravitational theory we already know, courtesy of Isaac Newton. All those spacetime-distorting effects that Einstein added only enter the picture close to the black hole's "event horizon"—the point where its gravity becomes so strong that even light can't escape from it (which is where the term "black hole" comes from). Since our spaceship would be thoroughly spaghettified long before it got this close, anything that happens at or near the event horizon is of strictly academic interest only.

Seen from a relatively safe vantage point outside the event horizon, a black hole is simply a very dense concentration of mass—rather like a neutron star, in fact, but compressed into an even smaller space. As long as we don't get too close it it (and aside from the fact that it's totally invisible), it might just as well be a neutron star in terms of the effect it has on us. This is a subtlety that Hollywood screenwriters often fail to grasp—there's really nothing exceptional in the way a black hole's gravity works, as long as you're a reasonably safe distance from it.

Let's think through a specific—and purely hypothetical—example. Suppose the Earth suddenly shrank down into a black hole (for the sake of simplicity, we'll picture this as happening without any giant explosion or any other non-gravitational effects). Surprising as it may seem, all the satellites moving in orbit around it would happily carry on as if nothing had happened. Astronauts on board the ISS might be perturbed by the fact they could no longer see the Earth, but they'd continue tracing out exactly the same orbit as before.

The fact is that *outside* the radius of the Earth, it doesn't matter—from a gravitational point of view—whether its mass is spread out in a sphere or if it's all concentrated in a single point at its centre. It's only when you look at what happens *inside* this radius that the situation changes. With the Earth as it really is, the pull of gravity towards its centre steadily gets weaker as you burrow down closer to it, eventually falling to zero at the very centre. On the other hand, if the Earth's mass was all concentrated into a single point—as in a black hole—then the gravitational attraction would just get stronger and stronger as you approached it.

So that's one Hollywood misconception out of the way. A distant black hole doesn't pull on a spacecraft any more strongly than a normal object of the same mass at the same distance. An even more egregious misconception is that a black hole can "suck" a passing spacecraft straight into it, like a vacuum cleaner. But that's emphatically not what would happen in reality. Instead, the

Fig. 2.10 In the real world, objects aren't "sucked" straight into a black hole—they break up into an accretion disc which gradually spirals inwards (NASA image)

spacecraft would simply be pulled into orbit around the black hole, just as it would around any other astronomical object. A black hole doesn't magically revoke the universe's conservation laws, so the spaceship has to retain whatever angular momentum it had to start with. Unless it had the bad luck to be on a head-on course for the black hole, it's simply going to swing around it—not fall straight into it.

Of course, objects can fall into black holes, but they do so very slowly. After tidal forces have pulled them apart, the debris spreads out into a so-called "accretion disc", and frictional forces inside this disc cause material to gradually lose angular momentum and spiral inwards (see Fig. 2.10).

If humans ever get close to a black hole, it's most likely to be a very small one, with a mass comparable to an asteroid rather than a star. One of the most realistic of all the early SF stories about black holes—and, perhaps as a consequence of its realism, not a particularly well-known one—is Isaac Asimov's "Old Fashioned", from 1976. This deals with a pair of astronauts whose spacecraft has an accidental encounter with a tiny black hole in the Solar System's asteroid belt—an unlikely but not impossible occurrence. Just as would happen in the real world, the black hole's effects on them are purely gravitational. As one of the pair says:

> Black holes can come in all sizes. That's what the astronomers say. That one is about the mass of a large asteroid, I think, and we're moving around it. How else could something we can't see be holding us in orbit?

In terms of their orbit around it, the black hole acts no differently from an asteroid of the same mass. What was different, however, was the jarring wrench that the black hole's steep tidal gradient gave them. As the same character goes on to say:

> We got close enough for the tidal effects to smash us up.... Even when the gravitational pull of a thing like that isn't large, you can get so close to it that the pull becomes intense. That intensity falls off so rapidly with increasing distance that the near end of an object is pulled far more strongly than the far end. The object is therefore stretched. The closer and bigger an object is, the worse the effect [44].

Another story from the same decade, dealing with a similarly tiny black hole, is Larry Niven's "The Hole Man" from 1973. While rather more far-fetched than Asimov's story—in effect, one of Niven's characters uses a microscopic black hole as a murder weapon—it still gets the basic physics right, in a way that a less meticulous writer might not have done. The victim, Childrey, doesn't die because he's "sucked" into the black hole. How could he be, when it only has the mass of a small mountain? Instead, as the murderer explains (after disingenuously claiming "I made a mistake"):

> I'd guess it massed about 10^{14} grams. That only makes it a millionth of an Angstrom across, much smaller than an atom. It wouldn't have absorbed much. The damage was done to Childrey by tidal effects as it passed through him [45].

References

1. A. Weir, *Project Hail Mary* (Kindle Edition), loc. 558
2. How Big Is Rosetta's Comet? European Space Agency, https://sci.esa.int/web/rosetta/-/54245-how-big-is-rosettas-comet
3. Three Touchdowns for Rosetta's Lander, European Space Agency, https://www.esa.int/Science_Exploration/Space_Science/Rosetta/Three_touchdowns_for_Rosetta_s_lander
4. A.C. Clarke, *2061: Odyssey Three* (Grafton Books, London, 1989), p. 90
5. A.C. Clarke, Summertime on Icarus, in *Tales of Ten Worlds*, (Gollancz, London, 2003), p. 21
6. A.C. Clarke, *1984 Spring* (Granada Publishing, London, 1984), pp. 149–150
7. A.C. Clarke, Hide and Seek, in *Of Time and Stars*, (Penguin Books, London, 1981), p. 114
8. A. Weir, *Project Hail Mary* (Kindle Edition), loc. 4202

9. H. Clement, Whirligig World, in *Astounding Science Fiction*, (June 1953), p. 108
10. F. Pratt, The Square-Cube Law, in *Thrilling Wonder Stories*, (1952), p. 93
11. L. Niven, The Borderland of sol, in *Tales of Known Space*, (Ballantine Books, New York, 1975), p. 176
12. I. Asimov, The Billiard Ball, in *Asimov's Mysteries*, (Panther Books, London, 1971), p. 249
13. A.C. Clarke, *2061: Odyssey Three* (Grafton Books, London, 1989), p. 68
14. A.C. Clarke, *A Space Odyssey*, vol 1968 (Arrow Books, London, 2001), p. 57
15. A.C. Clarke, *Rendezvous with Rama* (Ballantine Books, New York, 1974), p. 36
16. A.C. Clarke, *Rendezvous with Rama* (Ballantine Books, New York, 1974), p. 125
17. I. Asimov, For the Birds, in *The Winds of Change*, (Panther Books, London, 1984), p. 86
18. A. Weir, *Project Hail Mary* (Kindle Edition), loc. 4966
19. C. Hadfield, *The Apollo Murders* (Kindle Edition), loc. 2082
20. A.C. Clarke, Jupiter V, in *The Sentinel*, (Berkley Books, New York, 1986), pp. 162–163
21. A.C. Clarke, *The Other Side of the Sky* (Signet Books, New York, 1959), p. 29
22. Bergita and Urs Ganse, *The Spacefarer's Handbook* (Springer, Berlin, 2020), p. 84
23. L. Niven, J. Pournelle, *Lucifer's Hammer* (Futura Books, London, 1978), p. 50
24. A.C. Clarke, *The Hammer of God* (Orbit Books, London, 1994), pp. 211–212
25. I. Stewart, *Does God Play Dice?* (Penguin Books, London, 1989), p. 66
26. W. Ley, The Demotion of Pluto, in *Galaxy Science Fiction*, (August 1956), pp. 82–83
27. F. Hoyle, *The Black Cloud*, vol 35 (Penguin Books, London, 1960), pp. 35–238
28. A. May, *Cosmic Impact* (Icon Books, London, 2019), pp. 43–44
29. A.C. Clarke, *Astounding Days* (Gollancz, London, 1990), pp. 69–70
30. O. George, *Smith, the Complete Venus Equilateral* (Ballantine Books, New York, 1976), p. 11
31. Lagrange point, *The Encyclopedia of Science Fiction*, https://sf-encyclopedia.com/entry/lagrange_point
32. A. May, *The Space Business* (Icon Books, London, 2021), pp. 143–144
33. L. Niven, *Ringworld* (Gollancz, London, 2005), p. 56
34. L. Niven, *Neutron Star* (Orbit Books, London, 1978), pp. 14–15
35. A.C. Clarke, Neutron Tide, in *The Wind from the Sun*, (Pan Books, London, 1983), pp. 128–129
36. L. Niven, *Neutron Star* (Orbit Books, London, 1978), p. 28
37. C. Hadfield, *The Apollo Murders* (Kindle Edition), loc. 739
38. Tidally Locked Planet, *TV Tropes*, https://tvtropes.org/pmwiki/pmwiki.php/Main/TidallyLockedPlanet
39. L. Niven, *Introduction to Tales of Known Space* (Ballantine Books, New York, 1975), p. xii
40. L. Niven, The Borderland of sol, in *Tales of Known Space*, (Ballantine Books, New York, 1975), pp. 163–164

41. L. Niven, One Face, in *Convergent Series*, (Orbit Books, London, 1986), p. 31
42. Spaghettification, *Wikipedia*, https://en.wikipedia.org/wiki/Spaghettification
43. A. May, Supermassive Black Holes, https://www.space.com/supermassive-black-hole
44. I. Asimov, Old Fashioned, in *The Bicentennial Man*, (Panther Books, London, 1978), pp. 219–220
45. L. Niven, The Hole Man, in *A Hole in Space*, (Orbit Books, London, 1984), p. 143

3

Orbital Dynamics

The concept of an orbit is fundamental to the way that any object, whether artificial or natural, moves through space. Everyone knows that a satellite is in orbit around the Earth, but a spacecraft travelling from the Earth to Mars is also in orbit—although in this case, it's an orbit around the Sun rather than the Earth. This chapter describes the basic theory of orbits, showing how they arise naturally from Newton's law of gravitation. It also describes a range of phenomena associated with orbital dynamics that are sometimes encountered in well-informed science fiction, such as gravity wells, in-space rendezvous, Hohmann transfer orbits and gravitational slingshot manoeuvres, with examples drawn from the writings of Arthur C. Clarke, Larry Niven, Stephen Baxter, Andy Weir and others.

3.1 Why Orbits?

The concept of an orbit was introduced in the previous chapter on gravity, but it's important enough—and has enough ramifications—to deserve a more detailed chapter of its own. In essence, an orbit describes the motion of an object in a gravitational field when there are no other forces acting on it. In the situations we're concerned with in this book, the object in question would typically be a spacecraft, while the gravitational field acting on it could be that of a star, planet, asteroid or any combination of the above. The most important "other force" that could act on the spacecraft is the thrust of its rocket engine, which we'll only touch on briefly in this chapter in the context of changing from one orbit to another (the broader topic of "rocket science" is

A. May, *How Space Physics Really Works*, Science and Fiction,
https://doi.org/10.1007/978-3-031-33950-9_3

the subject of the next chapter). Between rocket burns, however, orbital theory is crucial to how a real-world spacecraft gets from A to B—and, on occasion, a science-fictional spacecraft too.

As we saw in the previous chapter, the form of an orbit in the simple case where only two objects are involved—such as a satellite orbiting the Earth, or the Earth orbiting the Sun—is very well established. In fact, there are many different ways in which the same solution can be reached. Historically, the first of these was the set of three empirical laws developed by Johannes Kepler early in the seventeenth century, followed by Newton's more general theory of gravity that came later in the same century (see Fig. 3.1). It's also possible to arrive at the same result by other means, such as the conservation of energy and angular momentum, which are arguably the most basic physical principles of all. At a more complicated level, it's possible to use Einstein's theory of General Relativity, which models the effect of gravity as a "curvature" of space and time.

But whichever of these approaches you use, when you apply them to the simple problem of one object orbiting in the gravitational field of another, you get the same answer; it traces out a specific type of oval-shaped path called

Fig. 3.1 A cartoon summarizing Isaac Newton's contribution to celestial mechanics, from an early science fiction magazine—the March 1940 issue of *Startling Stories* (Internet Archive)

an ellipse. If you picture orbits as commonly being circular, that's only because a circle is a special case of an ellipse that has zero eccentricity.

The previous chapter described the "thought experiment" of Newton's cannonball, showing how an object fired with sufficient horizontal speed from a high altitude will carry on forever in a circular orbit around the Earth. If, on the other hand, the object had slightly less than this "circular orbit" speed, it would dip down as it passed over the opposite side of the planet before rising up again. In other words, it would follow an elliptical orbit where the initial launch position was the highest point. Conversely, if the object was given *more* speed than it needed for a circular orbit, then it would travel in a larger ellipse where the launch point marked its closest approach to the Earth.

Not surprisingly, it requires mathematical calculations to work out the exact size and shape of an elliptical orbit for a given starting speed, but there's one further thing we can say even without going into such detail. Recalling from the first chapter that gravitational potential energy (PE) increases with increasing distance from the centre of attraction, then an elliptical orbit must have greater PE at its highest point than at the lowest point. The fact that energy is conserved then means that its kinetic energy (KE)—and hence its speed of travel—must follow the opposite trend, being fastest at the low point and slowest at the high point.

Unlike KE, which is just a function of speed and mass, PE is much less intuitive, and it may seem unnecessarily academic to talk about it here, in what is essentially a book about science fiction. Surprisingly, however, although the term itself is rare in SF, the basic notion of PE isn't. A fairly hackneyed concept in certain types of SF is that of a "gravity well", and this is basically just a way of talking about gravitational PE.

An early discussion of the idea can be found in a non-fiction book by Arthur C. Clarke from 1951, called *The Exploration of Space*. Clarke envisions the Earth's gravitational field as forming a deep, funnel-shaped well, with the Earth itself at the bottom (see Fig. 3.2). But this picture doesn't relate directly to something that would look like a well, or funnel, in the real world. It's easier to think of it as a kind of three-dimensional graph. The two horizontal dimensions correspond to the two real-world dimensions in the plane of a spacecraft's orbit, while the vertical axis represents something that can't actually be seen in the real world: the PE of the spacecraft at the corresponding location in its orbit. The funnel shape arises from the fact that PE increases with increasing distance from Earth.

Clarke sketched a number of possible orbits, of which (for simplicity) we've only included a few. The two labelled (f) or (g) are circular orbits—the latter having the greater energy of the two, and hence occurring further out from

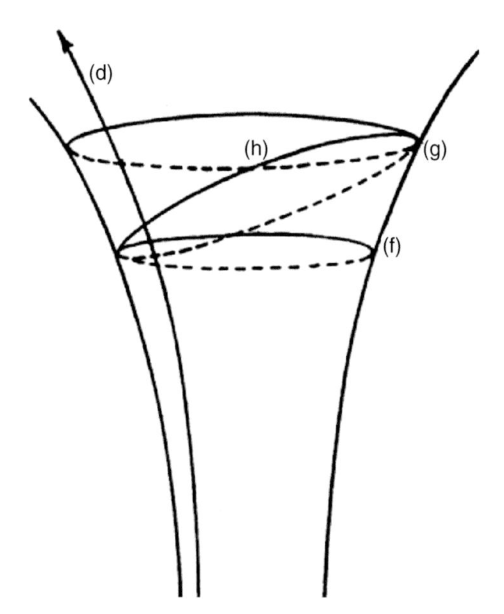

Fig. 3.2 A simplified copy of Arthur C. Clarke's sketch of the Earth's "gravity well"

the Earth. On both these orbits, the spacecraft's speed around the planet is just high enough that the resulting centrifugal force exactly balances the inward pull of gravity—hence it remains travelling in a perfect circle. But, as Clarke wrote:

> These are by no means the only possibilities that exist. Look, for example, at orbit (h). This is the path of a body which was projected horizontally at (g) but with insufficient speed to maintain itself. It fell downwards, gaining speed as it did so, until it had picked up enough speed to climb upwards again and retrace its orbit. Its path is thus not a circle, but an ellipse [1].

Conversely, if an object has too much speed for a circular orbit, it follows an ellipse which—rather than dropping down into the gravity well—slopes upwards and outwards. The higher the velocity, the more eccentric the ellipse—and as Clarke says:

> It is obvious that there will be one critical speed at which the body will never come back, but will creep over the top of the crater, as it were, and reach the horizontal level (d). This velocity is [11.2 kilometres per second] and is called "the velocity of escape". If a body started upwards with more than this critical speed, it would still retain the excess as a bonus when it escaped from the pit [1].

This, of course, is the important concept of "escape velocity", which has already featured in both the previous chapters. But attaining this velocity only means that you've climbed out of the gravity well of the planet you started on. You'll almost certainly still be within the much deeper gravity well of the star it revolves around—in our case, the Sun. You're still in orbit, but now it's a big orbit around a star rather than a smaller one around a planet. Moving from the influence of one to the influence of the other is a situation referred to by Stephen Baxter in his 1996 novel *Voyage*, when the spacecraft *Ares*, taking the first astronauts from Earth to Mars, receives the following message from mission control:

> You're now almost exactly 562,000 statute miles from the Earth. That's twice as far as any human has travelled before. And you're now passing out of the Earth's sphere of influence.

Baxter then goes on to define this "sphere of influence" more precisely:

> An imaginary bubble in space centred on Earth, an almost perfect sphere where the gravitational potential of Earth and Sun are in balance. Inside the sphere of influence, *Ares* had essentially been in an orbit dominated by Earth; beyond this point, however, the craft had escaped Earth and was in solar orbit [2].

Dotted around inside the Sun's gravity well are other smaller wells, created by the planets and other bodies that orbit around it. The Earth's gravity well, which we started with, is just one of these.

The larger a planet is, the deeper its gravity well—and consequently the higher its escape velocity. Since we spend our lives at the bottom of a relatively deep gravity well, we think of escape velocity as a major hurdle to be overcome before we can travel anywhere beyond our own planet. But this wouldn't be the case for any humans that might, at some point in the future, make their home in the asteroid belt. The bodies there are generally all small enough that their gravity wells pose no great obstacle, and travelling from one to another would require relatively little energy.

This is the premise underlying several of Larry Niven's early stories, in which he postulates a self-contained community of spacefaring "Belters". To them, gravity wells are essentially traps, which they refer to as "holes". In Niven's 1966 story, "At the Bottom of a Hole", the central character finds himself in the titular situation when he's forced to land on Mars. From his point of view, that's a very bad place to be. As one of the characters in the story says, "Belters learn to avoid gravity wells. A man can get killed half a dozen ways coming too close to a hole." [3].

The situation is described in a little more detail by SF writer Jerry Pournelle, in his non-fiction book *A Step Farther Out* from 1979:

> The Belters don't ever come to Earth or any other planet. Indeed, they regard planets as "holes", deep gravity wells which can trap them and use up their precious fuels. The assumption here is that it's far less costly to flit from asteroid to asteroid than it is to land on a planet [4].

For us down here on the Earth's surface, on the other hand, climbing up out of our gravity well and into orbit is the toughest part of the problem—even if we want to travel to the furthest reaches of the Solar System. Pournelle expresses this in a striking way by referring to Earth orbit as "halfway to anywhere". That's not literally true, of course, because Earth-orbiting satellites, some of them just a few hundred kilometres above the planet, are nowhere near halfway to any other Solar System destination. Mars, for example, is some 60 million kilometres away, even at its closest approach. But Pournelle wasn't thinking in terms of physical distance so much as energy requirements—and in this sense, Earth orbit really is pretty much "halfway to anywhere" [5].

In the next chapter, we'll see that a convenient way to discuss energy requirements in the context of rocket science is to use a quantity called "delta-v": the difference in velocity acquired by a spacecraft during a rocket burn. It happens that the minimum delta-v required to get from the Earth's surface into a useful orbit is around 8 kilometres per second—but is this really "halfway to anywhere"? Well, if you double that delta-v to 16 km/s, it could pretty much take you all the way to the dwarf planet Pluto, right on the edge of the Solar System, 6 billion kilometres away. So, yes—if we think simply in terms of the propulsive energy requirement, Pournelle was right, and Earth orbit really is halfway to anywhere (in the Solar System, at least).

There's a catch, though, in that any interplanetary journey still has to cover an enormous distance, which means that it's going to take a long time to complete. Just how long is something we'll come back to later in this chapter, when we get onto the subject of long-distance trajectories. Before then, however, it's worth taking a more careful look at a much more familiar kind of orbit.

3.2 Satellite Orbits

They say familiarity breeds contempt, and for many people today satellites represent the most boring aspect of space travel. They're used on a day-by-day basis for things as mundane as TV broadcasting and in-car navigation, all of which seem a far cry from the exciting space adventures of science fiction. Yet satellites have an intimate connection with the SF community in the person of Arthur C. Clarke, who played an important role in drawing attention to their utility for just the kind of everyday applications they're used for today.

Clarke's suggestion was originally published in the October 1945 issue of *Wireless World* magazine, in the form of a 4-page article entitled "Extraterrestrial Relays". The central idea—which wasn't new to Clarke, although pushing it as a viable commercial proposition was—revolved around the concept of "geosynchronous" satellites, of the kind predominantly used today for things like TV broadcasting. Orbiting at very high altitude, these appear to remain stationary over a fixed point on the Earth's equator—in contrast to satellites in low altitude orbits, which whiz by overhead fairly rapidly. At just 400 km altitude, for example, the International Space Station (ISS) only takes about 90 min to complete a whole orbit. This means that it revolves all the way around the Earth 16 times every day. Geosynchronous satellites, on the other hand, rotate at the same sedate rate as the Earth itself—which is why a TV satellite dish is always pointing in the right direction even though it never moves.

There are two separate reasons why satellites orbit more slowly the higher they are, and they're both fairly obvious. For one thing, higher-altitude satellites have further to travel, and for another, they feel a weaker pull of gravity—which means they need less centrifugal force, and hence less speed, in order to hold them in a circular orbit. As Clarke wrote in his 1945 article:

> A rocket which achieved a sufficiently great speed in flight outside the Earth's atmosphere would never return. This "orbital" velocity is 8 kilometres per second and a rocket which attained it would become an artificial satellite, circling the world forever with no expenditure of power—a second moon, in fact.... There are an infinite number of possible stable orbits, circular and elliptical, in which a rocket would remain if the initial conditions were correct. The velocity of 8 km/s applies only to the closest possible orbit, one just outside the atmosphere, and the period of revolution would be about 90 minutes. As the radius of the orbit increases the velocity decreases, since gravity is diminishing and less centrifugal force is needed to balance it.... It will be observed that one orbit, with a radius of 42,000 km, has a period of exactly 24 hours. A body in such an orbit, if its plane coincided with that of the Earth's equator, would revolve with

> the Earth and would thus be stationary above the same spot on the planet. It would remain fixed in the sky of a whole hemisphere and unlike all other heavenly bodies would neither rise nor set [6].

Clarke illustrates this with a rather technical-looking diagram (see Fig. 3.3). Note that he measures distances relative to the *centre* of the Earth, which gives his figure of 42,000 km for the geosynchronous orbit. This may appears at odds with the usually quoted figure of 36,000 km for "geosynchronous altitude", but the latter figure is measured relative to the Earth's *surface*—so it has the Earth's radius of approximately 6000 km subtracted from Clarke's value.

Interestingly, Clarke's work on geosynchronous orbits actually predates his first professional SF story, which appeared in the following year, 1946. While he's far better known among the general public for his fiction, and while there are other SF writers (such as Isaac Asimov and Gregory Benford) who had higher profile careers in the world of academic science, there's no disputing the historical significance of Clarke's *Wireless World* article. As Wikipedia says:

> The orbit, which Clarke first described as useful for broadcast and relay communications satellites, is sometimes called the Clarke Orbit. Similarly, the collection of artificial satellites in this orbit is known as the Clarke Belt [7].

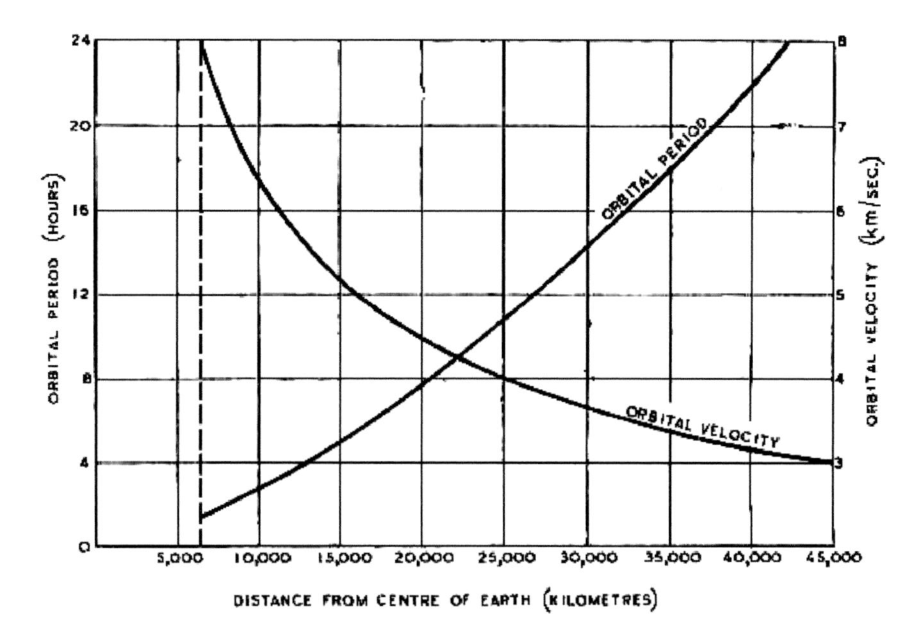

Fig. 3.3 A graph from Arthur C. Clarke's original 1945 article on geosynchronous satellites, showing how orbital time and speed vary with the size of the orbit (public domain image)

Sadly, having a region of space named after him was pretty much the only tangible reward Clarke received for his pioneering work. In his later career, he often bemoaned the fact that he hadn't tried to patent the idea, but instead simply sold the magazine article for a pittance—"not expecting that celestial mechanics would be commercialized in my lifetime", as he put it in his 1960 short story "I Remember Babylon" [8].

In 1966, 2 years after the first geosynchronous communication satellite was launched—and just two decades after his *Wireless World* article—Clarke wrote a piece gloomily subtitled "How I Lost a Billion Dollars in My Spare Time", in which he sums up the situation as follows:

> It is with somewhat mixed feelings that I can claim to have originated one of the most commercially valuable ideas of the 20th century, and to have sold it for just 40 dollars [9].

The importance of the "Clarke belt" can be see in Fig. 3.4, which shows the distribution of human-made objects—both satellites and space debris—around the Earth. The virtually solid white shell enveloping the planet shows just how densely populated the "low Earth orbit" region is, but when you

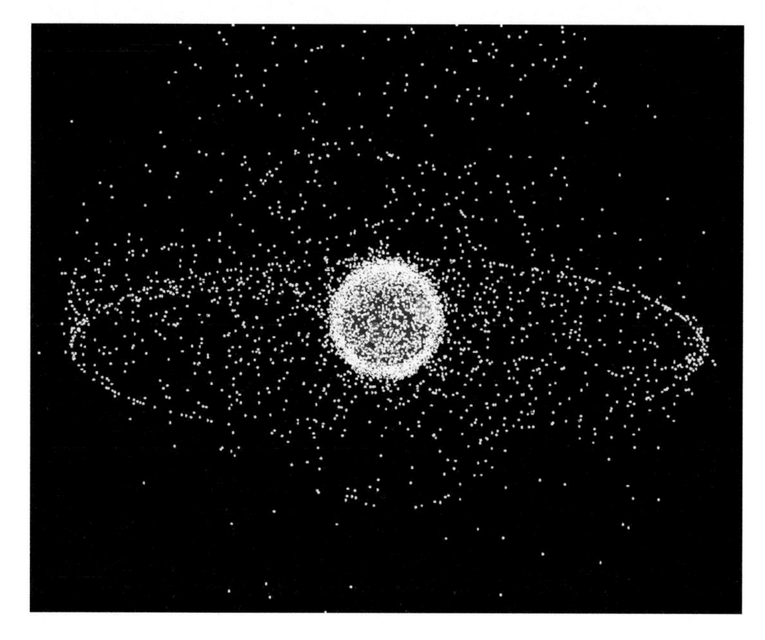

Fig. 3.4 This map of all the known objects in Earth orbit shows a high concentration in low Earth orbit, and a distinctive ring around the equator at the geosynchronous altitude (NASA image)

look beyond this, then the geosynchronous ring is the second most prominent feature.

At first sight, the existence of orbits gives the lie to the old saying "what goes up must come down". As we've seen, a satellite that's launched with sufficient speed will keep going round and round in orbit rather than falling back to the ground. In practice, however, this won't continue forever. Even at the altitude of most satellites, there's enough atmosphere that drag forces will slowly reduce a satellite's speed to the point that it will eventually fall back to Earth.

The higher a satellite is, the less air resistance it has to contend with, and hence the longer it will remain in orbit before coming back down to Earth. For a typical low Earth orbit, say at the 500 km altitude of the Hubble space telescope, a satellite should be able to stay up for around 20 years (the only reason Hubble has survived longer than this is that it's periodically been boosted back up to its working altitude). On the other hand, one of Arthur C. Clarke's geosynchronous satellites, at the much greater altitude of 36,000 km, won't fall out of the sky for hundreds of millions of years.

You sometimes see the "low Earth orbit" region defined as having a minimum altitude of 200 km. This lower limit isn't set by the physics of orbits *per se*—which in principle would allow an orbit at any altitude that isn't going to crash into terrain—but by the amount of air resistance a satellite would encounter. Occasionally space missions fail, not because they didn't reach orbit, but because the orbit was so low that it was effectively grazing the atmosphere. A high profile example occurred in 2019, during the first test flight of Boeing's Starliner capsule. It was meant to dock with the ISS at 400 km altitude, but in fact ended up in an elliptical orbit where the lowest point was a mere 187 km above the ground. In spaceflight terms, that's dangerously low, and the mission was scrubbed.

The idea of "orbital decay" due to severe drag forces is one that occasionally crops up in SF, as often as not in the context of a spacecraft orbiting another planet than the Earth. An example of this can be found in Andy Weir's 2021 novel *Project Hail Mary*, when the protagonist's ship finds itself in a perilously low orbit around an alien planet:

> Our orbit is still highly elliptical, and the closest part of it is way too close to the planet. Every 90 minutes, we touch the tippy-tippy-top of the atmosphere. It's barely an atmosphere at that altitude. Just a few confused air molecules bouncing around. But it's enough to slow us down just a teeny, tiny bit. That slowdown makes us dip a little deeper into the atmosphere on the next pass [10].

The hazards of atmospheric drag also play a part in Arthur C. Clarke's *2010: Odyssey Two* (1982), in the context of a spacecraft orbiting the Solar System's largest planet, the gas giant Jupiter. Here Clarke draws attention to another feature of such hazards, which is that—unlike the purely gravitational aspects of orbits, which are governed by relatively simple mathematics—drag forces can be extremely difficult to calculate. As a Russian character says to an American in the novel:

> Let me tactlessly remind you of an embarrassing incident from the old NASA days. Your first space station—Skylab. It was supposed to stay up at least a decade, but you didn't do your calculations right. The air drag in the ionosphere was badly underestimated, and it came down years ahead of schedule [11].

At the other extreme, there are bodies such as the Moon which, to all intents and purposes, lack an atmosphere altogether—and hence the problem of air resistance can be ignored. In this case, there's really no lower limit on the altitude at which an object can safely orbit, as long as it doesn't crash into any mountains. It's an idea that was used to comic effect by Ben Bova and Myron R. Lewis in their short story "Men of Good Will", from 1964. A brief but intense gun battle on the lunar surface creates a long-term problem in the form of endlessly orbiting bullets, as one of the story's characters explains:

> Only a few men got hit in the battle, none of them seriously. As in all battles, most of the rounds fired were clean misses. So one of our civilian mathematicians started doodling. We had several thousand very high velocity bullets fired off. In airless space. No friction, you see. And under low-gravity conditions.... They whizzed right along, skimmed over the mountain tops, thanks to the curvature of this damned short lunar horizon, and established themselves in rather eccentric satellite orbits [12].

It's a scenario that wouldn't quite work with today's military weapons, where a high-powered rifle might have a muzzle velocity approaching 1200 metres per second—distinctly less than the 1700 m/s needed for a circular orbit just above the Moon's surface. But the figures aren't that far adrift, so by the time humans visit the Moon in large enough numbers for serious quarrels to break out, orbiting bullets really might be a problem they need to watch out for.

3.3 Orbital Manoeuvres

When it's portrayed in movies, TV or video games, manoeuvring a spacecraft has a comforting familiarity. It's essentially the same as manoeuvring an aircraft in the Earth's atmosphere, or a submarine in the ocean. In the real world, however, things are very different. Spacecraft still obey the same laws of physics, but under such different conditions that the required approach is far from intuitive. For a start, the most non-intuitive thing of all—for the uninitiated, at least—is that an object's default state, when no forces are applied to it, is to remain in orbit. It doesn't slow down and grind to a halt, or plummet straight down to the Earth—it just keeps on going round and round. If that's what we want it to do, then all we have to do is leave it to its own devices. If not—for example if we want it to change its position slightly—it's only then that we need to take some action.

There are two fairly obvious reasons why a satellite might need to adjust its orbit: either to compensate for the effects of atmospheric drag, or to get out of the way of another satellite or large piece of space debris. In such cases the satellite can use relatively small rocket thrusters that allow it to make small changes in its speed or direction of motion. Another use of such thrusters is to alter a satellite's orientation in space, without actually changing the orbits it's moving on. This can be done using two offset thrusters pointing in opposite directions; firing them simultaneously for a brief time then causes the satellite to rotate through the desired angle. This might need to be done, for example, in order to keep the satellite's cameras, antennas and solar panels pointing in the correct direction.

There's another way to rotate a space vehicle, which makes direct use of the conservation of angular momentum. As mentioned in the first chapter, this is the basic principle that ensures that the total spin of a physical system always remains the same; if part of the system is set rotating in one direction, another part of it has to rotate in the opposite direction. This means that a suitable arrangement of flywheels inside the vehicle can be used to alter its orientation, by spinning the wheels in the *opposite* direction to the desired change in the vehicle itself. As low-tech as this procedure sounds, it does make at least one appearance in an SF context, when an engineer in Andy Weir's *Project Hail Mary* is explaining the mechanics of a small space probe:

> "How does it turn?"
>
> "Reaction wheels inside," he said. "It spins them one way, the ship turns the other." [13]

As long ago as 1951, in his non-fiction book *The Exploration of Space*, Arthur C. Clarke described how a flywheel could be used to stabilize a spacecraft that was tumbling end over end:

> Suppose there was a flywheel at the ship's centre of gravity, and imagine that at first it is fixed relative to the slowly turning ship. If now a motor starts to spin the flywheel … as the flywheel builds up speed the ship will gradually cease its rotation—until eventually it has come to rest and all the spin has been transferred to the flywheel. Since the flywheel will be very much smaller and lighter than the ship, it will obviously have to rotate at a very high speed to be effective [14].

One of the more familiar consequences of the conservation of angular momentum is the way that a spinning gyroscope always maintains its original orientation. This also has an application in spaceflight (and indeed, in the autopilots of terrestrial aircraft, missiles and torpedoes) as a way to maintain directional stability. It's now such a commonplace mechanism that it's rarely mentioned in SF, though back in 1951 gyroscopic stabilization was a novel enough concept that Arthur C. Clarke referred to it in *The Sands of Mars*. This was in the context of the spaceship *Ares*, travelling from Earth to Mars:

> The *Ares*, unlike the space station, was not turning on her axis but was held in the rigid reference system of her gyroscopes so that the stars were fixed and motionless in her skies [15].

A more exciting type of space manoeuvre—or one that has greater potential in the context of SF, anyway—is the rendezvous between one spacecraft and another. This is yet another topic where the physics is far from intuitive, as Bergita and Urs Ganse explain in their *Spacefarer's Handbook*:

> If a spacecraft were to launch into an orbit with the altitude and inclination of the International Space Station without any regards to its position on that orbit, the orbit times of the ship and the station would be identical…. They would simply circle around Earth indefinitely at a fixed distance from one another and would never rendezvous. It is likewise not a good idea to point a spacecraft right towards the intended target vessel, boost ahead, and expect to encounter it on a straight line. Since orbital motion occurs in ellipses, this would basically guarantee never to arrive at the intended destination. In science fiction depictions of spaceflight, this is one of the most common physics goofs! [16]

There's now a well-established technique for orbital rendezvous—and, whether you realize it or not, you've probably heard of the graduate physics student who developed it back in 1963. This was none other than Buzz Aldrin, who achieved enduring fame as the second human to step onto the Moon's surface in 1969. But 6 years prior to this, Aldrin had obtained his PhD degree for a thesis entitled "Line-of-Sight Guidance Techniques for Manned Orbital Rendezvous". The details are highly mathematical, but the basic principle boils down to something we've already seen in this chapter: objects in lower orbits move faster than objects in higher orbits. So the trick is for the chasing spacecraft to follow a slightly lower orbit than the one it's trying to rendezvous with, only switching to the higher orbit when it's finally caught up with it.

The technique was first demonstrated by NASA in 1965, when the Gemini 6 spacecraft successfully rendezvoused with—and approached within a metre of—its sister ship Gemini 7, which (despite its higher number) had been launched several days earlier (see Fig. 3.5).

Intercepting an orbiting spacecraft with another isn't simply a question of getting to the same point in space at the same time. The two vehicles also have to be travelling in exactly the same direction at exactly the same speed. This is the key to the final dramatic scene in Andy Weir's 2014 novel *The Martian*, which was reproduced—with the gratuitous addition of extra drama—in the movie of the same name starring Matt Damon as the protagonist Mark

Fig. 3.5 The Gemini 7 spacecraft photographed at close range from Gemini 6, after the two rendezvoused in orbit in December 1965 (NASA image)

Watney. The basic situation at this juncture is that Watney, stranded on Mars, has to launch himself into space using a small vehicle called the MAV, and then rendezvous with a fast-moving spacecraft called *Hermes* as it passes by overhead. Unfortunately, the MAV wasn't designed for this type of rendezvous, as a NASA engineer explains:

> The problem is the intercept velocity. The MAV is designed to get to low Mars orbit, which only requires 4.1 kilometres per second. But the *Hermes* flyby will be at 5.8 km/s [17].

So—since even people who haven't read the book or seen the film can safely assume the story has a happy ending—how does the MAV manage to attain a much higher speed than it was designed for? The solution is to strip its structure down to the bare minimum, thus drastically reducing its mass. Since its engine, like all rockets, is designed to give it a specific amount of momentum—and since momentum is mass times velocity—then a lower mass means it can reach a higher velocity.

Up to this point, we've been considering orbital dynamics as an essentially two-dimensional problem, with a spacecraft that's orbiting in one particular plane moving to another orbit—perhaps with a greater energy—in the same plane. This effectively involves taking the original elliptical orbit and simply expanding it into a larger ellipse. On the other hand, the problem becomes harder if the orbital plane itself has to change; in other words, taking the elliptical orbit and swivelling it up or down in the third dimension (see Fig. 3.6). If this has to be done by more than a few degrees, it can require a surprisingly large amount of energy.

The difficulty of switching from one orbital plane to another explains what may strike some people as a minor historical puzzle. When NASA discovered that the Space Shuttle *Columbia* had suffered serious damage during launch in 2003—serious enough, in the end, that it resulted in the complete loss of the Shuttle and its crew of 7 astronauts—why didn't they simply divert it to the ISS, where the astronauts could have awaited rescue? After the problem became public knowledge, a third of the way through the mission, this was one of the commonest "solutions" proposed by bloggers on the internet. Yet NASA's engineers never gave it any serious consideration at all, because they knew it was impossible. *Columbia* and the ISS were moving in completely different orbital planes, the former at an inclination of 39 degrees to the equator and the latter at 52 degrees. To change from one to the other would have involved a high-energy manoeuvre requiring far more fuel than the Shuttle was carrying [18].

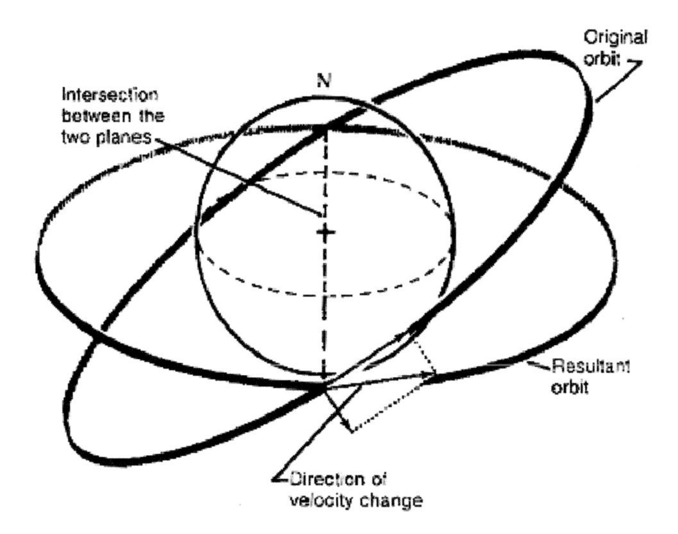

Fig. 3.6 The elliptical orbit of a spacecraft lies in a plane at a fixed inclination to the planet's equator. Changing from one orbital plane to another is far from trivial (NASA image)

So a rendezvous in Earth orbit—or around any other planet—can be hard enough. It's even harder, though, if you want two spacecraft to rendezvous at a random point in space, for example while en route between planets. It's a situation Arthur C. Clarke referred to in his novel *Earthlight*, from 1955:

> Spaceships travel at enormous velocities on exactly calculated paths, which do not permit of major alterations.... The orbit any ship follows from one planet to another is unique; no other vessel will ever follow the same path again, among the changing pattern of the planets. There are no "shipping lanes" in space, and it is rare indeed for one ship to pass within a million kilometres of another. Even when this does happen the difference of speed is almost always so great that contact is impossible [19].

This difference in speed also means that, in the real world, a high-speed flyby of an object such as a comet or asteroid is much easier to achieve than a genuine low-speed (or zero-speed) rendezvous. Even in the case of the most famous fictional rendezvous of all—Arthur C. Clarke's *Rendezvous with Rama* (1973)—the preliminary encounter with the eponymous Rama comes in the form of a fleetingly brief flyby by a robotic probe:

> There was no hope of a rendezvous; it would be the fastest flyby on record, for the two bodies would pass each other at 200,000 kilometres an hour. Rama

would be observed intensively for only a few minutes, and in real close-up for less than a second [20].

When we're talking about speeds, in the context of a rendezvous in space, we mean the *relative* speed between the two objects in question. This is another aspect of spaceflight that can be confusing for anyone not intimately familiar with it. Here on Earth, there's an obvious "standard of rest", in the form of the Earth's surface, and when we talk about speeds we always measure them relative to this standard. If you overtake a police car that's doing 30 miles an hour in a 30 mph speed limit, and you're doing 50 mph, it's no good trying to tell the officers you were only doing 20 mph relative to them. They're not physicists; they won't understand.

Looked at from a more fundamental point of view, however, you can never really talk about the speed of an object without qualifying it as being "relative to" something else. As Andy Weir puts it in *Project Hail Mary*:

> Velocity is relative. It doesn't make any sense unless you are comparing two objects. A car on the freeway might be going 70 miles per hour compared to the ground, but compared to the car next to it, it's moving almost 0 [21].

In a terrestrial context, this might sound like a pedantic quibble, but once you get out into space, relative speed is the only thing that really matters. It was irrelevant to Gemini 6 that Gemini 7 was orbiting the Earth at 28,000 kilometres per hour, because Gemini 6 was also orbiting at 28,000 km/h in the same direction; their *relative* speed was virtually zero. When Neil Armstrong and Buzz Aldrin were landing their lunar module on the Moon, they didn't care what its speed relative to the Earth was—they were only interested in its speed relative to the Moon's surface.

By the same token, you can never really claim to be stationary in space. Even if your speed is zero relative to the object you happen to be looking at, you—and it—are almost certainly going to be orbiting at high speed around something else. To quote Andy Weir from *Project Hail Mary* again:

> There is no "stationary" in a solar system. You're always moving around something [22].

Another consequence for space travellers en route between two locations is that, at any given moment, they will have one speed relative to their starting point and a *different* speed relative to their destination. As Arthur C. Clarke put it in his book *The Exploration of Space*:

> It is meaningless to speak simply of a spaceship's velocity without adding a little more information. For consider a ship between Earth and Mars. Observers on Mars, after a series of careful measurements, might decide that the ship was approaching them at 12,570 miles an hour. Equally accurate observations from Earth might show that it was receding at 8,490 miles an hour. Measurements by observers on the other planets would give different results again, since no two points in the Solar System are at rest with respect to each other. All the measurements would be equally "correct" [23].

This observation brings us neatly on to the subject of interplanetary travel—which, although it might not be obvious at first sight, is also all about orbits.

3.4 Interplanetary Trajectories

In Hollywood movies, the word "orbit" is usually reserved for the motion of a spacecraft around a planet, while actually travelling between planets—say from the Earth to Mars or vice versa—is presented as a completely different type of problem. In the latter case, you just point the spaceship in the required direction, fire up its rocket motors, and travel straight there—just like an airliner flying between London and New York. Unfortunately, the real world doesn't work like that. The Hollywood approach ignores two crucial facts: that both the Earth and Mars are themselves in constant motion around the Sun, and that the spacecraft will also feel the gravitational pull of the Sun and thus move in orbit around it.

To be perfectly fair to it, the Hollywood approach to Earth-Mars travel would work—but only if the ship's engines were so powerful that its speed, relative to the Sun, was much greater than that of the planets orbiting around it. As we'll see in the following chapter on rocket science, this isn't something that's ever going to be practical in the foreseeable future.

So let's try another idea—one that acknowledges that objects have to move in gravitationally feasible orbits around the Sun. Why not "hitchhike" most of the way to your destination on a passing comet or asteroid that happens to be going in the right direction? It's an idea with a venerable pedigree, having been used in one of the classic SF stories that Isaac Asimov chose to include in his retrospective anthology *Before the Golden Age* (1974). This was "Old Faithful", dating from 1934 and written by Raymond Z. Gallun. Asimov, however, didn't reprint the story because of its technical feasibility, but because he considered the title character—a Martian—to be one of first "sympathetic portraits of extraterrestrials" to appear in SF [24].

Like Asimov, Arthur C. Clarke was also an admirer of Gallun's story, as he recounts in his "science-fictional autobiography", *Astounding Days* from 1989. But as much as he praises "Old Faithful", he can't help pointing out its implausibility from the point of view of physics:

> I am sorry to say that Gallun helps to perpetuate a common astronomical fallacy. Old Faithful hitchhikes from Mars to Earth by flagging a ride on a convenient comet. Ignoring the fact that such miracles of timing involving three bodies must be extremely rare, the technique simply doesn't work [25].

The fundamental difficulty doesn't lie with the comet's trajectory—which might conceivably (in the spirit of the unlikely coincidences so often found in fiction) pass close to both Mars and Earth—but in the problem of rendezvousing with it. In fact, several years before he wrote the preceding words, Clarke had already debunked the idea of using a natural body—in this case an asteroid—as a free ride, in his non-fiction book *1984 Spring*:

> The fallacy arises, of course, from thinking of an asteroid as a kind of a bus or escalator. Any asteroid whose path took it close to Earth would be moving at a very high speed relative to us, so that a spaceship which tried to reach and actually land on it would need to use a great deal of fuel. And once it had matched speed with the asteroid it would follow the asteroid's orbit whether the asteroid was there or not [26].

Of course, it's perfectly possible for a spacecraft to rendezvous with a comet, as ESA's Philae probe demonstrated in 2014 when, as described in the previous chapter, it landed on comet 67P. From that point on, the spacecraft was effectively "hitching a ride" on the comet—but its trajectory through space was exactly the same as it would have been if the comet hadn't existed, and the probe had simply made the same set of preceding manoeuvres.

So, as far as the present context of orbital dynamics is concerned, hitching a ride on a comet or asteroid is a pretty pointless exercise. As an aside, however—and tangential to the subject at hand—there may in fact be other situations where it's not a bad idea at all. For example, if a robotic probe is exploring a series of asteroids, it may make sense for it to stay on one until it can hop over to another—possibly grabbing onto it with a tethered harpoon to avoid any expenditure of rocket fuel. As science-fictional as this may sound, it's a possibility that NASA is actually considering [27].

Returning to the problem of Earth-Mars travel, we can find a clue to the real solution if we look at an asteroid-type orbit a little more closely. One particular type of asteroid that was mentioned a couple of times in the previous chapter was the "Apollo" family, which travel in elliptical orbits where the innermost point is inside the Earth's orbit and the outermost point is outside Mars's orbit. Picture the extreme case, where the asteroid's orbit just grazes the Earth's orbit when it's closest to the Sun, and just grazes Mars's orbit at its furthest. This sounds like a very useful orbit to us—and remember that, because it's moving under the influence of the Sun's gravity, the asteroid doesn't expend any energy at all.

We've already seen that, as Arthur C. Clarke pointed out, there's no point hitchhiking on an asteroid because we'd expend the same amount of energy getting into its orbit as it would take to get onto that orbit even if the asteroid wasn't there. Now, however, this point works in our favour—because it means the asteroid doesn't actually need to exist at all. We just have to work out what orbit a useful asteroid *would* have, and then put our spaceship onto that orbit.

Here's how Clarke himself described the principle, in the same 1950s book, *The Exploration of Space*, that we've quoted from several times already:

> To move from an outer planet to an inner one means first of all losing speed slightly to drop down towards the Sun, then slowing down a little more as one goes past its orbit to match one's velocity with the inner planet. In exactly the same way, going to an outer planet means first increasing speed to climb upwards, then adding on a little more speed when passing the outer planet in order to keep up with it and avoid falling back again [28].

Except for the velocity changes at the start and end of trip, which require rocket burns, the ship will follow an asteroid-like elliptical orbit around the Sun without expending any effort at all. This means that, in the real world, travelling between planets may be less intuitive than the simple Hollywood approach of flying in a straight line, but it's also very much cheaper—because the spaceship's engines aren't needed most of the time. As long as the trip is planned carefully enough in advance, gravity will do most of the work.

The easiest path between two planets is an ellipse which is just so arranged that the inner planet lies at its innermost point and the outer planet at its outermost point. Counter-intuitively, this means that the best time to undertake an interplanetary mission is when the two planets are as far away from each other as possible, on opposite sides of the Sun (see Fig. 3.7).

A spacecraft trajectory of this type is called a "Hohmann transfer orbit", after the German scientist Walter Hohmann who first described it in 1925. It's by no means the only way to travel between two planets, but it's the one that requires the minimum amount of fuel. All the real-world Mars missions

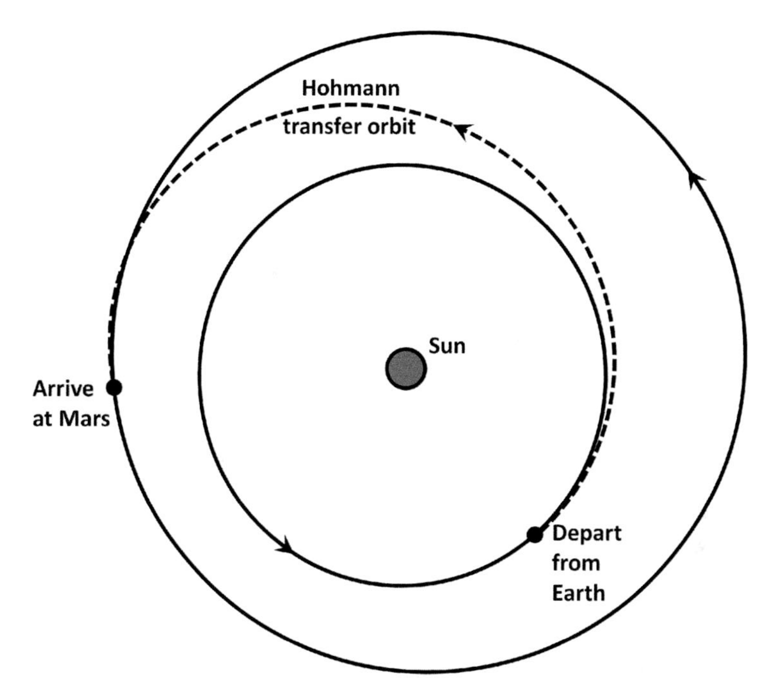

Fig. 3.7 The easiest path between two planets, such as the Earth and Mars as shown here, is a "Hohmann transfer orbit"—effectively a portion of a large elliptical orbit around the Sun

to date have used a Hohmann orbit, or something close to it—as, indeed, does the fictional Mars journey described in Stephen Baxter's novel *Voyage*. Written in 1996, this is set in an alternate version of the 1980s when NASA is imagined to have undertaken an Apollo-style mission to Mars. Here's how a NASA engineer in the novel describes the chosen route:

> Gregory walked to the blackboard and … drew two concentric circles on the board. "Here are the orbits of Earth and Mars. Every object in the Solar System follows an orbit around the Sun: ellipses, flattened circles, of one eccentricity or another. How are we to travel from Earth, on this inner track, to Mars, on the outer? We do not have the technology to fire our rockets for extended periods. We can only apply impulses, hopping from one elliptical path to another, as if jumping between moving streetcars. And so we must patch together our trajectory, to Mars and back, from fragments of ellipse. We kick and we coast; kick and coast…"

Gregory had drawn a half-ellipse which touched Earth's orbit at one extreme, and reached out to kiss Mars's orbit at the other. "Here we have a minimum-energy transfer orbit. It is called a Hohmann ellipse. Any other trajectory requires a greater expenditure of energy than this. To return to Earth, we must follow a similar half-ellipse." [29].

Even if the future brings improved rocket technology that allows faster journeys between planets, the Hohmann approach will still have advantages—for certain types of mission—because of its lower cost. This is something that Arthur C. Clarke alluded to in one of his earliest short stories, "Breaking Strain" from 1949:

> The *Star Queen* was 115 days on her orbit and had only 30 still to go. She was travelling, as did all freighters, on the long tangential ellipse kissing the orbits of Earth and Venus on opposite sides of the Sun. The fast liners could cut across from planet to planet at three times her speed—and ten times her fuel consumption—but she must plod along her predetermined track like a streetcar, taking 145 days, more or less, for each journey [30].

The Hohmann orbit does, of course, have one very obvious disadvantage; it only works when the departure and destination planets are in just the right parts of their orbit relative to each other. If you tried following a Hohmann trajectory, say from Earth to Mars, at any other time, then you would end up correctly at the orbit of Mars—but Mars itself would be in a completely different part of the the same orbit. This is why real Mars missions need to choose an appropriate "launch window" very carefully.

With robotic orbiters and landers, we're only interested in the one-way trip from Earth to Mars. Crewed missions, on the other hand, will have the additional consideration of the return journey. Here again, it will be essential to depart from Mars at just the right time to have any hope of getting back to Earth. It's a situation referred to by Gregory Benford in his 1999 novel *The Martian Race*:

> Orbital mechanics were clear and cruel. The whole round trip would take two and a half years. Due to the shifting alignment of the planets, launch windows for trajectories needing minimum fuel are spaced about 26 months apart. The trip each way takes about 6 months, leaving about one and a half years on the surface [31].

When Benford says "the trip each way takes about 6 months", there's no great mystery here. We've seen that the Hohmann orbit has to travel from one side of the Sun to the other (cf. Figure 3.7), and the Earth—of course—takes exactly 6 months to do that. In fact, because the spacecraft is travelling outwards to a greater distance from the Sun, it will actually take somewhat longer—about 6.7 months in the case of NASA's Perseverance mission, which landed on Mars in 2021.

So that's another disadvantage of the Hohmann trajectory: it takes quite a long time. This isn't so much because the spacecraft is travelling slowly, but because it's taking such a long, roundabout route to its destination. It's slow enough in the Earth-Mars case, but the further out from the Sun you go, the slower it gets. This casts a new light on something we touched on earlier in this chapter, in the context of Larry Niven's notion of a spacefaring "Belter" culture located in the asteroid belt. To quote what we said at the time: "the bodies there are generally all small enough that their gravity wells pose no great obstacle, and travelling from one to another would require relatively little energy."

Ah, but this is only true if the Belters use Hohmann-like orbits to travel between asteroids—and that's not going to be quick. In his book *A Step Farther Out*, Jerry Pournelle computed Hohmann launch windows and travel times between two locations in the asteroid belt. One was the dwarf planet Ceres, located about 2.7 astronomical units (AU) from the Sun—in other words, 2.7 times further out than the Earth—and the other was a hypothetical asteroid located at 2 AU from the Sun. The results were disappointing, to say the least, as Pournelle explains:

> Our belt civilization is in trouble again. Not only does it take 1.57 years to get from Ceres to 2 AU (or vice versa), but you can only do it once each 7 years! [32]

So Hohmann-type trajectories are never going to be practical for the routine and frequent interplanetary flights that are so often envisaged in SF. On the other hand, for the kind of one-off, carefully planned space missions that NASA undertakes today—and where rocket fuel is at a premium—they're indispensable.

In the case of ultra-long-distance trips inside the Solar System, there's another fuel-saving trick that can be used. If a close flyby of an intermediate planet can be worked into the trajectory, the very act of swinging around it can boost a spacecraft's velocity by means of a "gravitational slingshot" effect (see Fig. 3.8).

The beauty of this approach is that it really does provide "something for nothing". The spacecraft gains velocity in the slingshot encounter, not by using fuel or any other onboard power source, but by pinching a small amount of gravitational energy from the planet itself. It's a technique that's featured in a number of SF stories, perhaps most famously in Arthur C. Clarke's novel

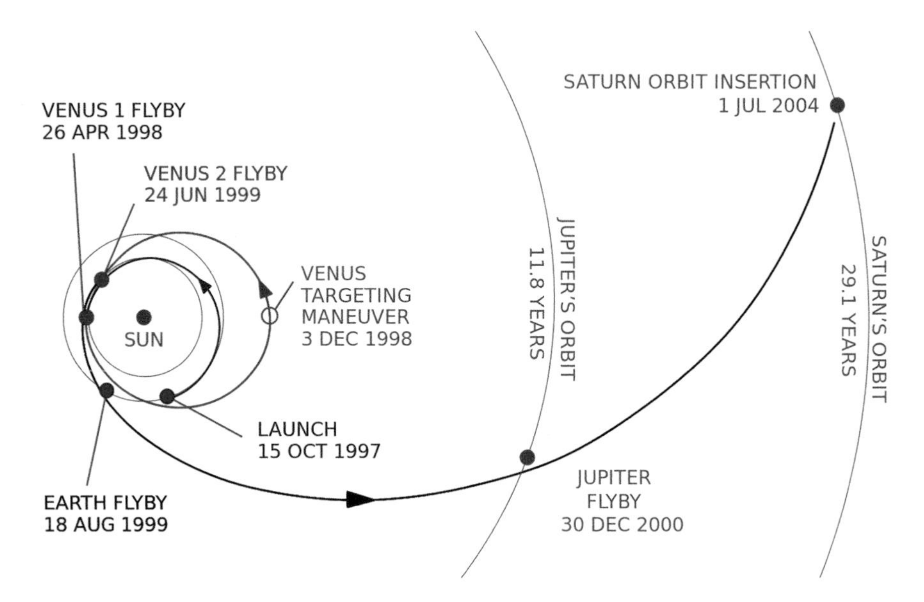

Fig. 3.8 NASA's Cassini mission took a roundabout route from the Earth to Saturn, using close encounters with Venus and Jupiter along the way to give it free energy boosts (NASA image)

2001: A Space Odyssey (1968). This differs from the movie version of *2001* in that the ultimate destination of the spaceship *Discovery* is Saturn, rather than Jupiter. However, it does have a close encounter with Jupiter to help it on its way:

> Like a ball on a cosmic pool table, *Discovery* had bounced off the moving gravitational field of Jupiter, and had gained momentum from the impact. Without using any fuel, she had increased her speed by several thousand miles an hour [33].

In *2010: Odyssey Two*, the sequel that Clarke wrote to *2001*, the spacecraft *Leonov* finds another way to exploit Jupiter's gravity to its own benefit. In this case, the spaceship's problem is escaping from orbit around Jupiter—and here's the solution that Clarke's astronauts come up with:

> We're going to use another trick, which—like so many of the concepts involved in space travel—at first sight seems to defy common sense. Although we're trying to get away from Jupiter, our first move is to get as close to it as we possibly can.... Our first burn ... will *reduce* our velocity, so that we fall down to Jupiter and just graze its atmosphere. Then, when we're at the closest possible point, we'll burn all our fuel as quickly as we can, to increase speed and inject *Leonov*

> into the orbit back to Earth. What's the point of such a crazy manoeuvre? It can't be justified except by highly complex mathematics, but I think the basic principle can be made fairly obvious. As we allow ourselves to fall into Jupiter's enormous gravity field, we'll gain velocity—and hence energy [34].

This quote talks rather glibly about "burning fuel"—a concept we've already used several times in this chapter, without actually explaining how it helps a spacecraft to gain speed or energy. That's an omission we'll rectify now, because the next chapter is all about rocket science.

References

1. A.C. Clarke, *The Exploration of Space* (Kindle Edition), loc. 550–60
2. S. Baxter, *Voyage* (Harper Collins, London, 2015), p. 159
3. L. Niven, At the Bottom of a Hole, in *Tales of Known Space*, (Ballantine Books, New York, 1975), p. 83
4. J. Pournelle, *A Step Farther Out* (Star Books, London, 1981), p. 32
5. J. Pournelle, *A Step Farther Out* (Star Books, London, 1981), pp. 22–25
6. A.C. Clarke, *Voices from the Sky* (Mayflower Books, London, 1969), p. 194
7. Geostationary orbit, *Wikipedia*, https://en.wikipedia.org/wiki/Geostationary_orbit
8. A.C. Clarke, I Remember Babylon, in *Tales of Ten Worlds*, (Gollancz, London, 2003), p. 2
9. A.C. Clarke, *Voices from the Sky* (Mayflower Books, London, 1969), p. 105
10. A. Weir, *Project Hail Mary* (Kindle edition), loc. 5949
11. A.C. Clarke, *Odyssey Two*, vol 1982 (Granada Publishing, London, 2010), p. 6
12. B. Bova, M.R. Lewis, Men of Good Will, in *Galaxy Science Fiction*, (June 1964), p. 174
13. A. Weir, *Project Hail Mary* (Kindle edition), loc. 4853
14. A.C. Clarke, *The Exploration of Space* (Kindle Edition), loc. 1150
15. A.C. Clarke, *The Sands of Mars* (Pan Books, London, 1964), p. 20
16. Bergita and Urs Ganse, *The Spacefarer's Handbook* (Springer, Berlin, 2020), p. 110
17. A. Weir, *The Martian* (Del Rey, New York, 2014), p. 327
18. S.K. Lewis, Rescue Scenarios, https://www.pbs.org/wgbh/nova/columbia/rescue.html
19. A.C. Clarke, *Earthlight* (Pan Books, London, 1966), p. 138
20. A.C. Clarke, *Rendezvous with Rama* (Ballantine Books, New York, 1974), p. 11
21. A. Weir, *Project Hail Mary* (Kindle edition), loc. 928
22. A. Weir, *Project Hail Mary* (Kindle edition), loc. 6386
23. A.C. Clarke, *The Exploration of Space* (Kindle Edition), loc. 1286
24. I. Asimov, *Before the Golden Age* (Black Cat, London, 1988), p. 567
25. A.C. Clarke, *Astounding Days* (Gollancz, London, 1990), p. 105

26. A.C. Clarke, *1984 Spring* (Granada Publishing, London, 1984), pp. 150–151
27. NASA, Comet Hitchhiker Would Take Tour of Small Bodies, https://www.nasa.gov/feature/jpl/comet-hitchhiker-would-take-tour-of-small-bodies
28. A.C. Clarke, *The Exploration of Space* (Kindle Edition), loc. 704
29. S. Baxter, Voyage (Harper Collins, London, 2015), pp. 72, 74
30. A.C. Clarke, Breaking Strain, in *The Sentinel*, (Berkley Books, New York, 1986), pp. 90–91
31. G. Benford, *The Martian Race* (Orbit Books, London, 2000), p. 25
32. J. Pournelle, *A Step Farther Out* (Star Books, London, 1981), p. 34
33. A.C. Clarke, *2001: A Space Odyssey* (Arrow Books, London, 1968), p. 128
34. A.C. Clarke, *Odyssey Two*, vol 1982 (Granada Publishing, London, 2010), p. 169

4

Rocket Science

In colloquial language, "rocket science" is virtually synonymous with "a very difficult subject". But it's really only the engineering details of rocketry that are intensely difficult, leading to its relatively slow rate of progress over the decades. The underlying science, on the other hand, is really very basic, having been well understood for more than a century. Nevertheless, those engineering difficulties mean that the space drives of science fiction—even when they're scientifically plausible—are generally far ahead of anything we have in the real world. From ion drives and nuclear thermal rockets to Bussard ramjets and relativistic time dilation, this chapter looks at some of the possibilities through the eyes of writers like Arthur C. Clarke, Andy Weir and Larry Niven.

4.1 Why Rockets?

Spacecraft need propulsion systems, though not in quite the same way that an aircraft flying in the Earth's atmosphere does. An aircraft needs continuous thrust simply to keep moving in the face of air resistance, but—with negligible air resistance in space—that's not true of a spacecraft. It only needs an engine at certain points, either when it has to increase its speed or change course. For all practical purposes, this engine will be some form of rocket—for reasons that will become clearer if we first consider how an aircraft's jet engine works inside the Earth's atmosphere.

At the front of the engine, there's a large-diameter intake that sucks in air as the aircraft flies along. Inside the engine, there's a combustion chamber that combines this air—or more specifically, the oxygen in it—with aviation fuel

A. May, *How Space Physics Really Works*, Science and Fiction,
https://doi.org/10.1007/978-3-031-33950-9_4

and burns it at high temperature. This creates an extremely hot gas, which—together with a large quantity of cooler, unburned air—is forced out at high pressure through a narrow-diameter exhaust nozzle at the rear of the engine. By the principle of conservation of momentum, this high-speed, backward-flowing exhaust transfers an equal and opposite forward momentum to the aircraft itself.

Obviously this isn't going to work in space, for the simple reason that there's no air for a jet engine to suck in. A rocket, on the other hand, is an engine that uses the same physical principle—the conservation of momentum—but works even in the absence of air. This means that, in addition to fuel, the rocket also has to carry its own oxidizer for combustion, while the combined propellant—the material that's eventually going to form the rocket's exhaust—has to have sufficient mass that when it's ejected the rocket will acquire the necessary amount of forward momentum.

The basic concept of a rocket is far from new, with simple ones based on gunpowder—a self-oxidizing solid propellant—being used in China as far back as the thirteenth century. However, they only became a serious practical proposition with the development of liquid fuelled rockets, equipped with turbo-chargers to create a sufficiently high-pressure exhaust, early in the twentieth century (see Fig. 4.1). By the middle of that century, although few people realized it at the time, rockets were ready to go into space.

Not surprisingly, the science fiction community was ahead of the field in grasping the significance of rocket travel. Although the basic concept predates that of the jet engine by centuries, the latter was the first to become an everyday reality. Observing that, as he put it, "the jet engine is, of course, a modified rocket engine, and operates on the same essential principles", John W. Campbell, the editor of *Astounding Science Fiction* magazine, wrote the following in April 1944:

> Since the Army announced the jet-propelled plane, many a science fiction author, reader—and editor—has discovered that friends, neighbours and acquaintances are abruptly beginning to believe that rocket ships aren't exclusively the province of wild fantasy, screwball inventors and impractical dreamers [1].

Around 6 months after Campbell wrote these words, Europeans woke up to the reality of rocket travel when—in the final months of World War Two—the first V-2 rockets were fired against targets in London, Antwerp and elsewhere. Designed by Wernher von Braun—who subsequently went on to develop the Juno rocket that put America's first satellite into orbit, and the

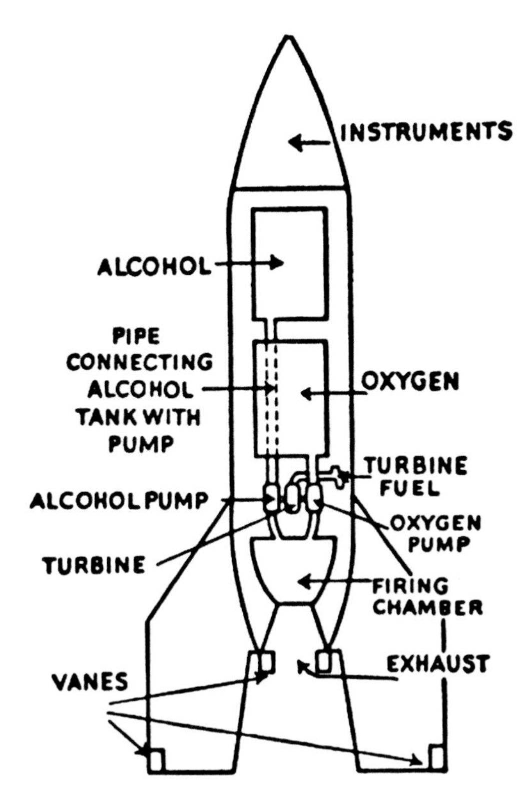

Fig. 4.1 Schematic diagram of a simple liquid-fuelled rocket, in this case using alcohol as the fuel (public domain image)

Saturn V that launched the Apollo missions to the Moon—the V-2 was perfectly capable of reaching outer space if fired vertically upwards. As another SF author, Robert A. Heinlein, has one of his characters observe in the 1950 novella *The Man Who Sold the Moon*, "The real engineering problems of space travel have been solved since World War Two." [2].

Not surprisingly, Arthur C. Clarke—one of the most technically astute SF writers of the period—was equally aware of the potential of rockets for space travel. He enumerated some of their basic characteristics in his 1951 non-fiction book *The Exploration of Space*:

1. The rocket will work in the absence of air or any other medium.
2. If the thrust or recoil remains constant, the acceleration will steadily increase as propellant is used up and the rocket becomes lighter.
3. The final speed depends directly on the ejection (or exhaust) speed: doubling the latter will double the former, and so on.

4. The final speed also depends on the weight of fuel ejected. If this is 1.72 times the final weight of the rocket, then a speed equal to that of the exhaust will be reached. If the ratio is 6.4 to 1, the rocket can travel twice as fast as its exhaust ... and so on [3].

The numerical figures that Clarke quotes come from the basic equations of rocket science, which are explained in greater detail by SF writer Jerry Pournelle in his non-fiction book *A Step Farther Out* (1979). To start with, Pournelle tells us that the force F pushing the rocket forward—in other words, its thrust—is given by the formula:

$$F = \left(M_0 - M_1\right) V_e / t$$

He goes on to explain this in plain English as follows:

> What this says is you take the mass of the ship when you get through burning fuel [M_1], subtract that from the mass you started with [M_0], divide by time of burn [t] and multiply by the exhaust velocity of the burned fuel [V_e], and you get thrust [4].

A second formula discussed by Pournelle is so important that it's often referred to as *the* rocket equation, with the definite article. This was originally derived as long ago as the nineteenth century by one of the great pioneers of rocket science, the Russian engineer Konstantin Tsiolkovsky. His famous equation relates the change in speed produced by a rocket burn—which, as mentioned in the previous chapter, is usually referred to as "delta-v"—to the speed of the rocket's exhaust and the ratio of its initial to final mass:

$$\text{delta-V} = V_e \ln\left(M_0 / M_1\right)$$

As Pournelle explains:

> What this says is that the total velocity change you can get from a rocket ship can be found by knowing the exhaust velocity of the fuel burned, and the ratio of the mass when you started to the mass when you finished the mission [4].

The only complication is that, rather than simply multiplying V_e by the ratio of the masses, you have to take the natural logarithm of the latter ratio. This is signified by "ln" in the formula, and it's a common enough mathematical function that it's usually found on most scientific calculators.

Because rockets can only carry a limited amount of propellant, they are generally only used for short "burns" lasting a few tens of minutes at most. Some advanced technologies do potentially allow more extended periods of thrust, and we'll look at a few of them later in this chapter—but to start with, we'll focus on the main task of present-day rockets: getting a spacecraft off the ground and into space in the first place.

4.2 Launch Rockets

When rockets switched from gunpowder-powered novelties to formidable liquid-fuelled vehicles like the V-2, it might have seemed like this was the end of the solid-fuelled rocket. Yet solid fuels—in far more efficient form than gunpowder—made something of a comeback after the start of the Space Age, and are now used alongside their liquid counterparts in today's launch rockets. The fact is, both types of fuel have their pros and cons.

One of the biggest advantages of a solid rocket is that it can be stored ready for use—and in perfect safety—for long periods of time, since the fuel and oxidizer are already mixed together, but won't burn until they're heated to high temperature by an igniter. This makes solid rockets particularly attractive for military applications, and the first really effective ones were developed in this context. By the early 1960s, long-range missiles such as the Minuteman, so named because it could be launched at a minute's notice, used purely solid propellants.

In the more peaceful context of space launchers, solid rockets have the additional advantage that they produce a greater amount of thrust for a given size and mass. On the other hand, liquid fuelled rockets remain the far more flexible option, insofar as they can be throttled up or down, or even switched off and restarted during flight. For this reason many of NASA's launch vehicles, such as the giant Space Launch System of the Artemis missions, or its predecessor the Space Shuttle, use a combination of both solid and liquid boosters to achieve the best of both worlds.

Solid rocket boosters—or SRBs—strapped to the sides of a giant liquid-fuelled launcher also feature in fictional form in Stephen Baxter's 1996 novel *Voyage*. Apart from the fictional setting, however, these SRB's behave just like real-world ones, as Baxter explains:

> The solid boosters were big firecrackers; once the SRBs were ignited, nothing could stop them until they burned out [5].

Firing up an SRB, once its internal igniter has been triggered, is a very quick business. It's quite a different matter, on the other hand, with a liquid-fuelled rocket. In the case of the main engines of the Saturn V rocket, for example, which launched the Apollo missions to the Moon, the process of ignition started a full 9 seconds before launch. This would have been obvious to anyone listening to the countdown of such a mission—although exactly what was happening during those 9 seconds might have remained a mystery. But the process is explained in detail by former astronaut Chris Hadfield, in his novel *The Apollo Murders* from 2021:

> "T minus 10, 9, and we have ignition sequence start."
>
> Four fireworks ignited inside each engine: two to spin up the fuel pump, and two to burn any flammable gases lurking in the exhaust nozzle.
>
> "6, 5, 4…"
>
> Two big valves opened, and liquid oxygen poured from its high tank down through the spinning pump and into the rocket, gushing out of the huge nozzle under its own weight like a frothy white waterfall. Two smaller valves clicked open, feeding oxygen and kerosene to fuel the jet engines, spinning the pumps up to high speed. The pressure in the main fuel lines suddenly jumped to 380 pounds per square inch. Conditions were set, with everything ready to ignite the rockets. Just needed some lighter fluid. Two small discs burst under the high fuel pressure, and a slug of triethyl boron/aluminium was pushed into the oxygen-rich rocket chambers. Like the ultimate spark plug, the fluids exploded on contact.
>
> "3, 2…"
>
> The middle engine lit first, followed quickly by the outer four; if all five had started at once, they would have torn the rocket ship and launch pad apart. Two more big valves opened, and high-pressure kerosene poured into the growing maelstrom….
>
> "One, zero, and lift-off, we have lift-off." [6]

In the course of this long drawn out sequence, a liquid-fuelled rocket has to build up enough thrust to get off the ground—and ideally it needs to do this as quickly as possible to avoid wasting valuable fuel while it's still on the launch pad. Impressively, exactly this consideration featured in an SF story even before the first satellites had been launched into orbit. The story, which appeared in the March 1955 issue of *Astounding Science Fiction* magazine, was called "The Test Stand", and it was written by a real-life rocket scientist named G. Harry Stine, using the pseudonym of "Lee Correy". Here is how Stine/Correy describes the situation:

> Today they don't think anything of igniting a whole rack of motors at the same time, but in those days we didn't know much about it … we considered ourselves mighty lucky to get one chamber going.… The motors would slobber around and waste a couple tons of propellants—liquid oxygen and liquid hydrogen—before the system built up enough thrust to get off the ground. Then some bright design engineer back in the plant got the brilliant idea that those tons of fuel could be saved if the motors started at full thrust [7].

Once it's off the ground, a rocket has to work hardest in the lower atmosphere where the pressure and density of the air are at their highest. The total effective pressure resisting the rocket's upward thrust is the sum of the normal—or "static"—air pressure P and a second component called the "dynamic" pressure, or Q. The latter is equal to half the local air density multiplied by the square of the rocket's speed. This means that Q initially increases, as the rocket speeds up, then eventually starts to decrease again as the air density drops. In other words, at some point Q reaches a maximum value—as many people will be aware if they've ever watched a space launch, since the commentator often refers to the point of "max Q".

One author who has experienced max Q for himself is Chris Hadfield, who of course achieved fame as an astronaut before he wrote his novel *The Apollo Murders*. Given the hyper-realism of that novel, it's no surprise that it makes reference to max Q:

> The air had been pushing back against their increasing speed, putting a heavier and heavier load on the structure of the ship. The engineering name for that pressure was Q, and they had just hit the speed and height where it was at maximum: max Q. From now on, with the air getting rapidly thinner with height, the forces would drop. It was the heaviest load the rocket's structure was designed for [8].

As we saw earlier in this chapter, a rocket has to carry all its propellant with it from the start, unlike a jet aircraft that can draw in large quantities of air from the atmosphere. As a result, a rocket vehicle will have far more mass than a jet that produces the same amount of thrust. In fact, a rocket scientist's greatest headache is the constant battle between the thrust generated and the mass it has to push.

We saw one consequence of this in the previous chapter, in the context of *The Martian*—both Andy Weir's original novel and the Hollywood film made from it—when the Mars ascent vehicle had to be drastically lightened in order to reach the desired orbit. There's a similar, though less extreme, example in Chris Hadfield's *The Apollo Murders*—in this case, also drawing on the problem of altering a spacecraft's orbital plane, which we also encountered in the previous chapter.

Hadfield's novel features the launch of a Saturn V rocket that was originally designed to inject its payload into a near-equatorial orbit, flying more or less directly east from the launch site in Florida. But it's decided at a late stage of the proceedings that the fictional Apollo 18 has to rendezvous with another spacecraft in a more highly inclined orbit. This calls for a completely different flight path—"After launch, you'll need to steer up the Florida coast," as one of the characters says. The flight director's reply is blunt and straight to the point: "Can't do it.… If we don't launch the Saturn V straight east out of Canaveral, we don't get the added speed of the Earth's spin." The only solution—as in the case of the MAV in *The Martian*—is to reduce the mass of the payload by removing everything that isn't absolutely essential [9].

The need to achieve the best possible balance between the thrust and mass of a launch rocket leads to one of the most characteristic features of all present-day launchers: the multi-stage design. The basic idea is simple enough; when all the propellant in a tank has been used up, the empty tank becomes an unnecessary dead weight—so why not get rid of it? Yet as obvious as this seems, it's a relatively rare concept in the world of SF. One striking exception is Isaac Asimov's short story "The Martian Way"—which is even more remarkable for having been published in 1952, long before multi-stage rockets became familiar to the general public. Here is how one of the story's characters explains the basic principle:

> Now imagine a spaceship that weighs 100,000 tons lifting off Earth. To do that, something else must move downward. Since a spaceship is extremely heavy, a great deal of material must be moved downwards. So much material, in fact, that there is no place to keep it all aboard ship. A special compartment must be built behind the ship to hold it.… But now the total weight of the ship is much greater. You will need still more propulsion and still more.… When the material inside the biggest shell is used up, the shell is detached. It's thrown away too.… Then the second one goes, and then, if the trip is a long one, the last is ejected [10].

Despite Asimov's use of words like "compartment" and "shell" to refer to the rocket stages, this is a perfectly good description of the way a multi-stage rocket works today (see Fig. 4.2). It implicitly draws on the principle of the conservation of momentum, when it says the propellant must move downwards in order for the spaceship itself to move upward. It also makes clear one of the basic realities of rocket science, that the mass that finally reaches orbit is only a small fraction of the mass that is initially lifted off the ground.

In their *Spacefarer's Handbook*, Bergita and Urs Ganse use Tsiolkovsky's rocket equation to show that staging isn't just a fuel-saving luxury, but an essential requirement if you want to launch a payload into orbit. As we saw in the previous chapter, this requires a delta-v of at least 8 km/s—or even more if, as in the case of the Apollo missions, you then want to go on somewhere else, such as the Moon. In the real world, the Apollo spacecraft was launched by the three-stage Saturn V rocket, but let's see what would have happened if it had consisted of just a single stage containing exactly the same amount of propellant.

Fig. 4.2 Illustration of the upper stage of a launch rocket—in this case a SpaceX Falcon 9—separating from the first stage after the latter has burned all its fuel (NASA image)

According to the Ganses, the Saturn V stack had a mass of 2970 tonnes when it was fully loaded with fuel, and just 261 tonnes when it was empty. For this first calculation, we're assuming that all that mass was concentrated in a single stage, so we can just put those figures straight into Tsiolkovsky's equation. The resulting delta-v is a mere 6.1 km/s—clearly nowhere near enough to reach orbit, let alone the Moon. Now let's look at the real-world situation, where the same amount of fuel was divided up between three stages, with each stage being jettisoned as soon as it was out of fuel. The same equation, applied three times in succession, now yields a much more acceptable final delta-v of 10.2 km/s [11].

The Saturn V, like most launch rockets, was strictly single-use, with each stage being permanently discarded once it was empty. Now, however, we're beginning to see a new generation of reusable launch vehicles, such the SpaceX Falcon 9—only the first stage of which is reusable—and the same company's Starship, which is intended to be fully reusable. The latter consists of two stages, the lower one returning to the launch site after completing its job, while the upper stage continues on to Earth orbit.

Virtually the same launch procedure was anticipated by Arthur C. Clarke in the book version of *2001: A Space Odyssey* (1968). Viewers of the movie only see the Orion shuttle after it's reached orbit and is approaching the space station—but Clarke's novel describes the whole flight, starting with a launch from the Kennedy Space Centre in Florida:

> In a 10,000 mile arc the empty lower stage would glide down into the atmosphere, trading speed for distance as it homed on Kennedy. In a few hours, serviced and refuelled, it would be ready again to lift another companion [12].

That's almost exactly how SpaceX intends a Starship mission to play out. Another trick that SpaceX is planning for Starship is the ability to refuel the upper stage in Earth orbit—and again, this is something that was anticipated in fiction by Arthur C. Clarke. In this case, the relevant passage comes from his early novel *Prelude to Space*, from as far back as 1953:

> Beta could leave the atmosphere, but she could never escape completely from Earth. Her task was twofold. First, she had to carry up fuel tanks into orbit round the Earth, and set them circling like tiny moons until they were needed. Not until this had been done would she lift Alpha into space. The smaller ship would then fuel up in free orbit from the waiting tanks, fire its motors to break away from Earth, and make the journey to the Moon [13].

The laborious-sounding process that Clarke describes here—with multiple launches of the same vehicle stockpiling fuel in orbit, before finally launching the crewed spacecraft—is essentially the same procedure that SpaceX is planning for lunar missions using Starship. As the *Teslarati* website reported in 2021:

> First, SpaceX will launch a custom variant of Starship that was … confirmed by NASA to be a propellant storage (or depot) ship…. Second … the company would begin a series of 14 tanker launches spread over almost six months—each of which would dock with the depot and gradually fill its tanks. Third, once the depot ship is topped off, the actual Starship Moon lander would launch, dock with the depot, and be fully fuelled [14].

4.3 Other Ways to Reach Space

There's no doubt that present-day space launchers yield some impressive statistics, whether it's NASA's super-powerful Space Launch System with a total thrust exceeding that of 30 large airliners, or SpaceX's super-efficient Falcon 9 with an average launch frequency of better than one per week. For all that, however, there's something rather clunkily old-fashioned about the whole idea of multi-stage rocket launchers, and—probably for this reason—they rarely feature in SF visions of the future.

Of course, SF is perfectly capable of breaking the laws of physics, whether it's though miraculous anti-gravity drives or simply ignoring the need to conserve energy and momentum. Here in the real world, however, are there any other space launch options besides a rocket?

At the most practical end of the spectrum, one possibility has been implicit in the discussion ever since the start of this chapter. We saw that a jet aircraft doesn't need to carry its raw propellant along with it, because it can suck in as much air as it needs from the atmosphere as it goes along. It still needs to carry fuel, of course, but that's only a small fraction of the total mass that's expelled in pushing the jet along. A rocket, on the other hand, needs to carry a lot more mass than a jet, because it can't collect any air when it's in space.

That's the way the argument usually goes, but it ignores a crucial point. For a large part of the time that a rocket is being launched, it isn't in space at all—it's still inside the atmosphere. So, while it's there, why does it have to be a rocket at all—why can't it simply be an ordinary "air-breathing" jet?

This is the thinking behind the Skylon launch concept, currently under development by a British firm called Reaction Engines Limited. In effect,

Skylon is a hybrid spaceplane that combines both a jet and a rocket in one package. When it's ascending through the dense lower parts of the atmosphere—or descending back down again—it works just like a conventional jet, drawing in air from the atmosphere to provide both oxygen for combustion and exhaust mass to push itself along. However, after reaching an altitude of around 30 km, where the air is too thin for the engine to work efficiently as a jet, it switches into rocket mode, using its own onboard supply of oxygen [15].

Yet again, this is another idea that Arthur C. Clarke anticipated long ago in *Prelude to Space*. The fictional space launcher in that novel happens to be nuclear-powered—something that's perfectly feasible in the real world, as we'll see later in this chapter—but it still needs to eject some form of propellant mass in order to push itself along. Just like the Skylon concept, it does this using atmospheric air whenever possible, as Clarke explains:

> As long as her air-scoops could collect and compress the tenuous gas of the upper atmosphere, the white-hot furnace … would blast it out of the jets. Only when at last the air was too thin for power or support need she … become a pure rocket [13].

As it happens, we can stick with Arthur C. Clarke for a couple of other launch methods, which—although they sound like pure science fiction—are equally possible in the real world. The first makes an appearance in his 1952 novel *Islands in the Sky*, where a character refers to it as "that magnetic thing that shoots fuel tanks up to rockets orbiting the Moon." Here is a more detailed description of the contraption in question:

> It's an electromagnetic track about 5 miles long, running east and west across the crater Hipparchus. They chose that spot because it's near the centre of the Moon's disc, and the fuel refineries aren't far away. Ships waiting to be refuelled get into an orbit round the Moon, and at the right time they shoot up the containers into the same orbit [16].

What Clarke is describing here is an electromagnetic concept that goes by the rather vague name of a "mass driver." This differs from a rocket in that the energy source—and the launcher itself—remains fixed on the Moon's surface, so the object being launched can be virtually pure payload. The concept is explained in more detail by science writer Brian Clegg as follows:

> A mass driver uses a series of electromagnets to accelerate a payload along a track to such a speed that it can be catapulted into space.... To escape the Moon without subsequent thrust being applied, a ship would have to be accelerated to around 2.3 kilometres per second. This sounds a lot, but it could be achieved with a 27 km long track, providing an acceleration of 10 g—just about acceptable for humans [17].

You'll notice that Clegg, like Clarke, places his launcher on the Moon rather than the Earth. This is because the Moon has a much weaker gravity field than the Earth—and no air resistance to overcome—so the acceleration needed can be kept down to manageable levels. In fact, in a lunar context, the approach could well have major advantages over a rocket launcher (being basically electrical in nature, it could easily be solar powered, for example). As science-fictional as an "electromagnetic space launcher" sounds, it's an idea that has been given serious consideration by NASA for a future lunar base (see Fig. 4.3).

The most audacious of all the space launch options considered by Clarke is the "space elevator", which provides the central focus of his 1979 novel *The Fountains of Paradise*. As a theoretical concept, the space elevator arises as a natural extension of the geosynchronous orbit we encountered in the previous chapter—and, as we saw there, has a strong connection with Clarke himself, even if he didn't originate the idea.

To recap what we learned earlier, the existence of the geosynchronous orbit means that a satellite placed in a circular orbit at 36,000 km over the Earth's equator will appear to remain stationary above a fixed point on the Earth's surface. A hypothetical "space elevator" would then run from this fixed point all the way up to the geosynchronous altitude, where any object released into space would automatically be in orbit itself.

In his novel, Clarke attributes the idea to a Russian engineer named Yuri Artsutanov "in the very decade that the first satellite was launched"—in other words, the 1950s. As Clarke's protagonist explains, Artsutanov's reasoning went as follows:

> If the laws of celestial mechanics make it possible for an object to stay fixed in the sky, might it not be possible to lower a cable down to the surface, and so to establish an elevator system linking Earth to space? [18]

In an afterword to *The Fountains of Paradise*, Clarke confirms that Artsutanov was a real person, and also notes that the same idea was subsequently discussed by a group of American researchers in a piece entitled "Satellite

Fig. 4.3 NASA visualization of how an electromagnetic space launcher, or "mass driver", might look on the Moon (NASA image)

Elongation into a True Skyhook", which appeared in *Science* magazine in 1966 [19].

The construction of a space elevator, or skyhook, is described in some detail by Robert L. Forward in his 1995 book *Indistinguishable from Magic*. This is a book that combines both science fiction and non-fiction, and this particular piece falls in the latter category (with Forward writing in his professional capacity as an aerospace engineer):

> A Skyhook would be built from the middle out, starting with a cable-making machine at a Central Station in geostationary orbit. For balance, the machine would extrude two cables, one upward and one downward. The cables would be thin at first, then, when the length of the cable hanging down became longer, the thickness of the cable would have to be increased to provide enough strength to support the increasing weight below. The thickness of the upward-growing cable would also have to increase as the cable became longer, but for a different reason. Instead of the Earth gravity pulling on the cable, the pull is due to the centrifugal force from the once per day rotation about the Earth.

> If the extrusion rates of the two cables are carefully controlled, then the net pull on the Central Station in geostationary orbit would be zero, and the cable laying machine would remain in geostationary orbit. Eventually, the lower end of the cable would reach the ground (or the top of some convenient near-equatorial mountain) 36,000 kilometres below. At that time, the outgoing cable would be 110,000 kilometres long. The outgoing cable has to be longer than the Earth-reaching cable because of the way the gravity forces and centrifugal forces vary with distance [20].

With dimensions thousands of times bigger than any current engineering construction, it's clear that building a space elevator would be an extraordinarily ambitious undertaking. It also calls for advanced technologies that don't exist yet—particularly relating to the material used for the ultra-long cable itself—and it poses hair-raising issues of safety and logistics. For all these reasons and more, space elevators are probably never going to be a practical proposition in the real world. Nevertheless, they remain a perfectly valid theoretical possibility, so you shouldn't sneer too much when you occasionally encounter them in SF.

4.4 Advanced Space Propulsion

Once they're off the ground and in space, most present-day spacecraft only use their rocket engines for relatively brief "burns", whenever they need to change from one orbit to another. Perhaps the most famous instances of this were the "TLI" and "LOI" burns during the Apollo missions to the Moon. These are explained by Chris Hadfield in *The Apollo Murders* (in the context of a fictitious Apollo mission with the callsign *Pursuit*) as follows:

> TLI ... Trans-Lunar Injection, firing the rocket motor that accelerated them to escape velocity out of Earth orbit, headed for the Moon.

and

> LOI ... Lunar Orbit Insertion, the firing of *Pursuit*'s main rocket engine to slow the ship down, allowing it to be captured by the Moon's gravity into a stable orbit [21].

In the real world, space missions are traditionally designed in such a way that such rocket burns are kept to a minimum, in order to reduce the mass of fuel that needs to be carried. This is at odds with the situation commonly depicted in SF, where spaceships often keep their engines running all the time, resulting in a continuous steady acceleration (or deceleration, as they approach their destination). Is this mode of working pure fantasy, or are there really ways that a spacecraft could run its engines for an extended period of time?

The answer to that is yes, technologies do exist that can produce continuous acceleration in the real world—but they come with a catch. You can never get something for nothing (except in fiction), and if you want your acceleration to be continuous then it's going to be a very gradual one.

The most mature of these technologies, which has been used by NASA on several of its small interplanetary probes, is "solar electric propulsion"—colloquially known by the more science-fictional sounding name of an "ion drive". This works on the same physical principle—the conservation of momentum—as an ordinary rocket motor, but it has a key difference. In a rocket, the energy comes from a fuel-oxidizer mixture that's carried on board, and which also serves as propellant mass. In an ion drive, on the other hand, these two functions—energy supply and propellant—are completely separate.

As the official name "solar electric propulsion" suggests, the energy for an ion drive comes from the spacecraft's solar panels—and hence it's virtually limitless. But the resulting electrical power can't be used in the same way it would in a terrestrial vehicle—by turning wheels or a propeller, for example—because there's no ground surface or air flow for them to work against. Instead, the electricity is used to produce a super-fast stream of ionized gas, which is then ejected from the spacecraft to propel it along in the same way as a conventional rocket exhaust.

One of the highest profile uses of an ion drive to date was in NASA's Dawn spacecraft (see Fig. 4.4), which visited two different destinations, Vesta and Ceres, in the asteroid belt. In NASA's own words:

> Ion propulsion systems ionize (charge) atoms and then exploit their non-neutral charge to expel them from the spacecraft, creating thrust. Dawn's gridded ion thrusters achieve this by accelerating xenon, ionized by an electron beam, through a voltage between two charged grids.... Ion propulsion requires less propellant than traditional chemical systems, so ion propulsion systems add less mass to the spacecraft, making them far more efficient [22].

The secret to an ion drive's efficiency is its super-high exhaust velocity, which can be up to 10 times faster than that of a conventional chemical rocket. The downside, however, is that current designs have to use their xenon propellant so sparingly that the resulting acceleration is incredibly

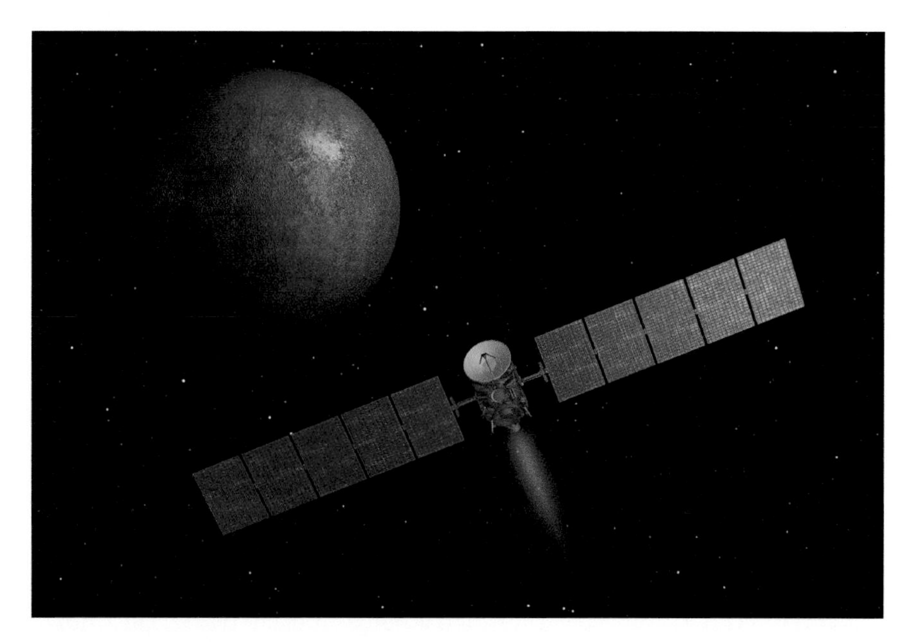

Fig. 4.4 Artist's impression of the Dawn spacecraft, with its ion drive thrusting it along, as it arrived at Ceres in 2015 (NASA image)

tiny—around 7.5 millionths of a "g" (the acceleration due to the Earth's gravity) in the case of Dawn. But this acceleration can be kept up all the time, thus allowing the spacecraft to build up a substantial delta-v over the course of an entire mission. In Dawn's case, this was a very respectable 6.7 km/s, which compares favourably with the delta-v of a typical Apollo TLI burn, which was under 5 km/s.

In science fiction, the term "ion drive" has been a familiar piece of technobabble since the 1940s—albeit used, more often than not, in such a vague way that it is simply a synonym of "futuristic propulsion system". However, one SF author who certainly did know what the term meant was Andy Weir, when he used it in his 2014 novel *The Martian*, in the context of the *Hermes* spacecraft carrying a 6-person crew between the Earth and Mars:

> Gone are the days of heavy chemical fuel burns and trans-Mars injection orbits. *Hermes* is powered by ion engines. They throw argon out the back of the ship really fast to get a tiny amount of acceleration. The thing is, it doesn't take much reactant mass, so a little argon (and a nuclear reactor to power things) let us accelerate constantly the whole way there. You'd be amazed at how fast you can go with a tiny acceleration over a long time [23].

From this description, which appears at the start of the book, you could imagine that Weir is talking about a scaled-up version of Dawn's ion drive, using a nuclear reactor instead of solar panels to generate electricity, and argon as opposed to xenon as the propellant (though the two gases are so similar it hardly makes any difference). However, a throwaway remark by one of the characters later in the novel—"I repaired the bad cable on VASIMR 4"—suggests that actually Weir had a somewhat different type of "ion engine" in mind [24].

That cryptic word "VASIMR" happens to be the trademarked name of a spacecraft engine that the Texas-based Ad Astra Rocket Company has been working on since the early 2000s. It stands for "Variable Specific Impulse Magnetoplasma Rocket", and it's comparable to a Dawn-style ion thruster in that it uses electricity as its main energy source, and it involves ionising a gas. However, rather than using a stream of positively charged ions as the propellant, it creates a neutral plasma that contains both positive ions and negatively charged electrons. It then channels this plasma through a strong magnetic field and ejects it at high speed through a flared, rocket-style exhaust nozzle. This produces a greater amount of thrust than a standard ion drive (though still less than a traditional rocket), and it can likewise be kept up for long periods of time. According to Ad Astra, a nuclear-powered VASIMR engine could get a crewed spaceship to Mars in under 40 days, far faster than a present-day missions [25].

If we've got a nuclear reactor on our spaceship, there's another, more straightforward way it can be used to power it. Why not simply use the reactor to heat a gas—any gas—to a high enough temperature that it can be used like a conventional rocket propellant? This is the idea behind a fictional spaceship that's already been mentioned a couple of times in this chapter—the one in Arthur C. Clarke's novel *Prelude to Space*, from 1953. The basic rationale for using nuclear power is almost a no-brainer, as Clarke says:

> Atomic reactions … are a million or more times as powerful as chemical ones…. The energy released by the few pounds of matter in the first atomic bombs could have taken a thousand tons to the Moon and back.

He then goes on to explain how this energy can be transformed into propulsive thrust:

> In the chemical rocket, we get our driving exhaust by burning a fuel and letting the hot gases acquire speed by expanding through a nozzle. In other words, we exchange heat for velocity—the hotter our combustion chamber, the faster the

> jet will leave it. We'd get the same result if we didn't actually burn the fuel at all … we could make a rocket by pumping any gas we liked—even air—into a heating unit, and then letting it expand through a nozzle [26].

Although the word "nuclear" has many negative connotations, one thing that no one can dispute is that it can pack a large amount of usable energy into a small volume. Just a few kilograms of uranium, for example, can keep a 15,000-tonne nuclear submarine going for a year. So, if a submarine can be powered by a nuclear reactor, why not a spaceship?

The simplest approach is exactly the one described by Clarke back in 1953, which goes by the generic name of a "nuclear thermal rocket". In fact, that's a perfect description of the principle: "nuclear" because the energy comes from a nuclear reaction, "thermal" because this energy is released in the form of heat, and "rocket" because the heat is used to create a fast-moving stream of super-hot exhaust gas.

In the real world, NASA pursued plans for a nuclear thermal rocket to an advanced stage during the Apollo era. Called NERVA—for "Nuclear Engine for Rocket Vehicle Application"—it would have served as a more powerful replacement for the third stage of the Saturn V launch vehicle. With twice the exhaust velocity of the original third stage, and a longer duration burn amounting to 30 min or so, it would have resulted in a much higher delta-v—probably enough to get all the way to Mars. Unfortunately (or fortunately, if you dislike all things nuclear) the project was cancelled by the US Congress in 1972 and never made it into space.

Even so, NERVA was an entirely viable proposition from a technical point of view (see Fig. 4.5). Had it gone ahead, it might have made certain missions possible in the 1970s or 80 s—such as a crewed journey to Mars and back—that, as things turned out, can't be accomplished even today. In a fictional context, NERVA, or something very like it, makes guest appearances in this role in both Gregory Benford's *The Martian Race* [27] and Stephen Baxter's *Voyage* [28].

So, even today, we have designs like Dawn's ion drive that can produce tiny amounts of thrust over a period of several years, and NERVA—on paper, at least—which can produce a substantial amount of thrust for half an hour or so. Given this, it's not beyond the bounds of possibility that some future technology might offer the best of both worlds—a substantial thrust for an extended period of time.

One potential future breakthrough that might lead in this direction is in the area of nuclear fusion, which is a very different process from the nuclear fission used in present-day nuclear reactors. Fusion produces even more

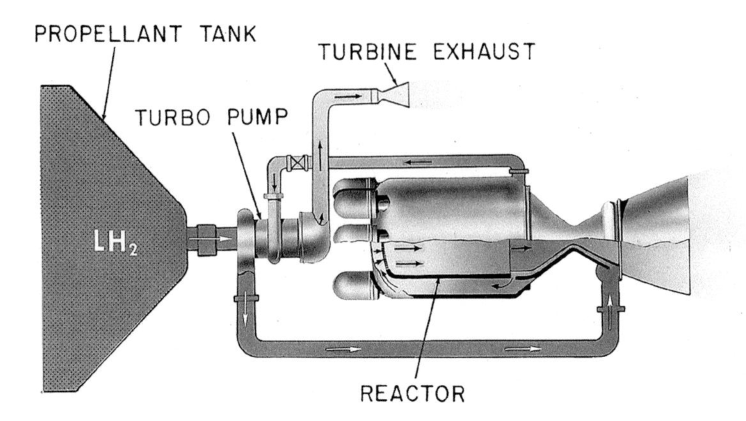

Fig. 4.5 Schematic diagram of the NERVA nuclear thermal rocket concept (NASA image)

energy per kilogram than fission, and it uses a much more convenient fuel—hydrogen, which is ubiquitous in the universe, as opposed to scarce elements like uranium or plutonium. It's possible that a fusion-based propulsion system could yield the ultimate goal of a large amount of thrust sustained for a long period of time, thus coming much closer than current technology to the kind of space travel traditionally depicted in SF. As one NASA engineer put it: "Interplanetary missions would no longer need to wait for a shortest-journey launch window. You can launch when you want." [29].

In the case of a crewed mission, there's an interesting consequence of continuous-thrust propulsion that we came across earlier, in the chapter about gravity. We saw there how astronauts feel "weightless" inside a spacecraft when its engines are switched off, but experience a kind of "artificial gravity" when the engines are producing thrust. The strength of this effect can be related directly to the Earth's own gravity, if the spaceship's acceleration is measured in multiples of g (i.e. 9.8 m/s^2, which is the acceleration due to gravity at the Earth's surface).

As we've seen, the kind of continuous acceleration that's possible with current technology is such a tiny fraction of g that it wouldn't be perceptible to astronauts inside the spaceship. But what about the future, when it might be possible to accelerate continuously at a rate of, say, 1 g? That's probably the maximum sustained acceleration that that would be comfortable for a human crew without special training. Not only would they feel the same familiar pull of "gravity" as on Earth, but—with their speed increasing by almost 10 m/s every second—they would get to their destination amazingly quickly. In fact, interplanetary travel times would finally come down to the kind of level typically portrayed in SF.

Actually, it's easy enough to estimate such travel times using a few simple calculations. Jerry Pournelle did just this in his non-fiction book *A Step Farther Out*, where he calculated the time that various journeys would take at a constant acceleration of 1 g. His results are quite an eye-opener. You could cover the distance from the Earth to the Moon in 3.5 h, to Mars or Venus in around 3 days, to Saturn or any of its moons in 8 days, and all the way out to Pluto in just 20 days [30].

In practice, of course, you don't literally want to accelerate all the way to your destination or you would overshoot it at high speed. The best procedure would be to accelerate at 1 g as far as the half-way point, then turn the ship round so that its engines are pointing forwards rather than backwards, allowing you to decelerate at 1 g for the remainder of the journey to your destination. This is the procedure that Arthur C. Clarke describes in his 1975 novel *Imperial Earth*, in the context of a trip from Saturn's moon Titan to the Earth:

> With the advent of constant acceleration drives which could maintain thrust for the entire duration of a voyage, midpoint or "turnaround" had a real physical meaning.... They could watch the slow rotation of the stars as the ship was swung through 180 degrees, and the drive was aimed precisely against its previous line of thrust to whittle away slowly the enormous velocity built up over the preceding 10 days [31].

Of course, we only chose a 1 g acceleration for reasons of comfort; if there was any cause for urgency, the acceleration could be even higher than this. We saw an example of this in an earlier chapter, when the spaceship in Andy Weir's *Project Hail Mary*—on an emergency mission to save the Earth—accelerates to its destination at 1.5 g. Another example occurs in Larry Niven's *World of Ptavvs* from 1965. Here the protagonist chooses to accelerate at a distinctly unpleasant 2 g, in order to catch up with two other ships that have a head start of a day and a half—but are only accelerating at a standard rate of 1 g. He calculates that it will take him 2 days to overtake them, on the basis that "given continuous acceleration, the decrease in trip time varies as the square root of the increase in power." [32].

4.5 Interstellar Travel

In the real world to date, even the very longest space journeys—such as that of NASA's New Horizons probe to Pluto and beyond—are limited to the confines of our own Solar System. Travel to other star systems is still purely the domain of science fiction. This, however, is only a matter of practicality—not of theoretical possibility. The main obstacles to an interstellar spaceflight are the enormously long, probably multi-century, journey times they would involve—which of course is a major deterrent to any government or private company embarking on one—and the huge energy requirements, both for propulsion and in-flight support systems. On the other hand, from a purely theoretical point of view, there are various technologies within the known laws of physics that would make such a journey possible.

The biggest problem with a rocket engine—even a nuclear powered one like VASIMR or NERVA—is that it only provides thrust as long as its on-board propellant supply lasts. Once this runs out, the spacecraft will just coast along at whatever speed it's achieved up to this point. More practically, the crew would be well advised to cease forward thrust when half the propellant is gone, to leave enough to decelerate again when they reach their destination.

It would be much more convenient if the ship could collect fresh propellant as it travels along, in much the same way that an aircraft's jet engine collects air from the Earth's atmosphere. If space was a complete vacuum this would be impossible, but as it happens it's never completely empty. Even in the space between stars, there are a few atoms of hydrogen, at the incredibly low density of around one atom per cubic metre. Back in 1960, the physicist Robert Bussard proposed that this ultra-tenuous gas might serve both as an energy source and propellant mass for an interstellar spaceship.

The concept of a "Bussard ramjet" (see Fig. 4.6) appealed to one SF author in particular, Larry Niven, who used it in several of his early stories. Here is how he described the principle in a non-fiction essay entitled "Bigger than Worlds", from 1974:

> A Bussard ramjet would use an electromagnetic field to scoop up the interstellar hydrogen ahead of it—with an intake a thousand miles or more in diameter—compress it, and burn it as fuel for a fusion drive [33].

In his more familiar role as a writer of fiction, Niven employed the idea as the central plot driver of his short story "Rammer", from 1971, which was subsequently incorporated into his 1976 novel *A World out of Time*. Here's a colourful passage from it:

> A rammer was the pilot of a starship. The starships were Bussard ramjets. They caught interstellar hydrogen in immaterial nets of electromagnetic force, compressed and guided it into a ring of pinched force fields, and there burned it in

Fig. 4.6 A NASA artist's impression of what a Bussard ramjet might look like in flight (NASA image)

> fusion fire. Potentially there was no limit at all on the speed of a Bussard ramjet. The ships were enormously powerful, enormously complex, and enormously expensive [34].

Some people might query Niven's statement that there is "no limit" on a Bussard ramjet's speed, given the common understanding that material objects can't exceed the speed of light. However, as we'll see shortly, the situation isn't as clear-cut as it's often made out to be—it really all depends on how you define "speed"—and there's a sense in which Niven's statement is perfectly true.

As science-fictional space drives go, the Bussard ramjet has the huge advantage that it doesn't break any fundamental laws of physics, and it's based on the entirely sound reasoning that interstellar hydrogen atoms could be used as fuel for a nuclear fusion reactor. But does this mean it would be possible to build one in the real world? Sadly, the answer is probably not. A detailed study carried out in 2021 concluded that the associated engineering challenges are likely to prove insurmountable—with, for example, the magnetic "ram scoop" needing to be at least 4000 km in diameter [35].

As far as present-day technology is concerned, there's only one way that humans could send a spacecraft—and only a miniature robotic one at that—to a nearby star in a meaningful timeframe. It's called a "laser-propelled light-sail", but before discussing it in detail it's worth an aside on the closely related—but simpler—concept of a solar sail.

This is primarily a book about space physics, and we've seen multiple times already that one of the most fundamental principles in this field is the conservation of momentum. This means that an isolated, self-contained spacecraft can only gain momentum if it projects an equal amount of momentum—for example a rocket exhaust or a stream of ionized gas—in the opposite direction. On the other hand, if the spacecraft isn't an isolated, self-contained system, there's another way it can gain momentum: by absorbing it from an external source. The best analogy here is with the way that a sailing ship on Earth is pushed along by the wind. What's actually happening is that the blustery gas molecules in the atmosphere are transferring some of their momentum to the ship's sails.

Is there an analogous wind in space? Actually there is, although it's something of a red herring in the present context. We'll see in the next chapter that the Sun throws out streams of high-energy particles in the form of a "solar wind", but as it turns out this isn't the thing that's exploited by solar sails. Instead, they depend on the physical pressure exerted by sunlight itself, which constitutes "free momentum" in the same way that the Sun's light falling on a solar panel can be considered "free energy".

Here is how Arthur C. Clarke—or rather, his protagonist—describes the principle in the short story "The Wind from the Sun", from 1964:

> Hold your hands out to the Sun ... what do you feel? Heat, of course. But there's pressure as well—though you've never noticed it, because it's so tiny. Over the area of your hands, it comes to only about a millionth of an ounce. But out in space, even a pressure as small as that can be important, for it's acting all the time, hour after hour, day after day. Unlike rocket fuel, it's free and unlimited. If we want to, we can use it. We can build sails to catch the radiation blowing from the Sun.

As with Dawn's ion drive, the rate at which a solar sail accelerates would be tiny, but it could be kept up over a very long time. As Clarke's protagonist goes on:

> Its acceleration will be ... about a thousandth of a g. That doesn't seem much, but let's see what it means. It means that in the first second, we'll move about a fifth of an inch. I suppose a healthy snail could do better than that. But after a minute, we've covered 60 feet, and will be doing just over a mile an hour. That's not bad, for something driven by pure sunlight! After an hour, we're 40 miles from our starting point, and will be moving at 80 miles an hour ... [and] at the end of a day's run, almost 2,000 miles an hour. If it starts from orbit—as it has to, of course—it can reach escape velocity in a couple of days. And all without burning a single drop of fuel! [36]

Clarke's story envisions fairly hefty solar-sail powered craft being used by human crews for recreational purposes. However, with the miniaturization of electronic components that's possible today, it's more likely that solar sails will be put to a rather different use, comparable to the 10-cm "cubesats" that have started to proliferate in Earth orbit. Such miniaturization allows the payload part of a solar-sail craft to be very small and light, so the sail itself—an ultra-thin sheet of highly reflective fabric—doesn't need to be impractically huge.

One such small-scale vehicle, the Planetary Society's LightSail 2 demonstrator, has already been flown successfully in space. Launched in June 2019, this was basically a three-unit cubesat—in other words, it measured 10 by 10 by 30 cm—which deployed a 5.6 m square Mylar sail after it was placed in Earth orbit. It then used this sail, rather than the rocket thrusters of a more conventional satellite, to move itself from one orbit to another—all courtesy of the free momentum given to it by sunlight (see Fig. 4.7).

Now that we know all about solar sails, we can return to the topic we broke off from a couple of pages ago: laser-propelled light-sails. This idea can be traced back to the 1970s, and an aerospace engineer that we met previously in this chapter—Robert L. Forward, who doubled up later in his career as an SF writer as well. Forward's idea was to propel a light-sail equipped spacecraft, not with the diffuse light from the Sun, but with a narrowly focused, high-energy beam directed straight at it from a purpose-built installation. In modern treatments of the idea, this installation is usually a super-powerful laser, but Forward's original conception used a microwave beam instead.

Here is a description of the essential elements of the set-up by Forward himself—not from his original technical proposal, but from a later SF story that he wrote around the idea called "Fading into Blackness" (1988):

> First was the design for the large solar power satellite, ten times larger than any that had ever been designed before. It would generate the 50,000 megawatts of

Fig. 4.7 An artist's impression of LightSail 2 in Earth orbit (Planetary Society, CC-BY-SA-3.0)

> microwave power that would be needed to accelerate the interstellar probe up to 20 per cent of the speed of light.
>
> Second was the design for the radar dish that was to focus the microwaves into a far-reaching beam. It was ultra-large, four times larger than the diameter of the Earth, and ultra-light, made out of a thin disc of fine wire mesh. The wire mesh disc was to be shaped into a nearly perfect parabola by centrifugal acceleration from a slow rotation of the disc combined with a linear acceleration from a million electric ion thrusters spaced around the periphery of the disc.
>
> Finally, there was the interstellar probe itself, a wisp of ultra-fine wire woven into a hexagonal mesh that was a kilometre in diameter, yet had a total mass of only 16 grams. At each six-wire intersection of the mesh was a microcircuit, so tiny that all 100 billion of the microcircuits had a total mass of only four grams [37].

Some sceptics might roll their eyes at Forward's use of the word "interstellar", dismissing it as typical science-fictional exaggeration. But the word is genuinely appropriate in this context, and in fact a system considerably less ambitious than the one Forward described might enable miniature space probes to reach another star system within a human lifetime. This, anyway, is the gist of a proposal put forward by the Russian billionaire Yuri Milner in 2016.

Called Breakthrough Starshot, Milner's proposal involves sending a large fleet of tiny spacecraft—potentially several hundred of them—to the Alpha Centauri system, 40 trillion kilometres away. Each craft would be equipped with a sail similar in size to that of LightSail 2, while being much smaller in mass—just a few grams each. They would be accelerated on their way using an array of ground-based lasers—carefully pointed in the direction of Alpha Centauri—with a combined power of 100 gigawatts. According to Milner's calculations, the laser array would be capable of accelerating each Starshot craft to 60,000 km/s—a fifth of the speed of light—in just 10 min of continuous operation.

The reason Milner talks about such a large fleet of craft is to hedge his bets, because it's likely that many of them will miss their target, or break down en route. But he's hopeful that at least a few of them would get close enough to catch a glimpse of the Alpha Centauri system as they whiz past (there's no way to slow them down when they get there, of course). Even with their tiny size, it should be possible with today's technology to cram in enough equipment to take a few photographs, and then send them back to Earth using tiny onboard laser transmitters.

At the speed Milner envisages the Starshot craft as travelling, it should only take them 20 years to travel the 4 light-years to Alpha Centauri—followed by another 4 years for the transmitted data to get back to Earth. Given that a more conventional space probe might take something like 70,000 years to cover the same distance, that's a pretty impressive feat—and it really does look to be possible with technology that exists today [38].

Of course, being able to send miniaturized probes on a 20-year journey between two stars that are actually very close together, on a galactic scale, isn't much use to SF writers. For virtually any story that's going to grab a reader's attention, you need to be able to send a crewed spaceship much further than that in a much shorter time. As a result, most fictional interstellar spaceships have, of necessity, to break one or more of the currently understood laws of physics. Occasionally this is done blatantly, using a term like "reactionless drive"—which is a euphemism for one that doesn't obey the conservation of momentum—but more often than not, authors simply ignore the subject of physics altogether.

Plenty of entertaining books could be written—and indeed have been written—about speculative future developments in real-world physics that might enable easy interstellar travel that resembles its traditional portrayal in SF. The present book, on the other hand—being restricted to the laws of physics as we know and understand them today—will have to skip over such speculative considerations. But there's one aspect of futuristic space travel that *does* fall within the remit of currently understood physics, and this concerns the far from intuitive behaviour of objects that travel close to the speed of light. As a subject, this is rather more advanced than most of the physics in this book, but even so it's intriguing enough to be worth taking a brief look at.

4.6 The Problem of Relativity

When contemplating travel—or indeed communication—over the kind of distances involved in space flight, we inevitably come up against that notorious speed limit: "nothing can travel faster than light". But where does this limit come from, and what does it really mean? The answer lies in a theory developed in 1905 by Albert Einstein (see Fig. 4.8), called "special relativity".

The first thing to say about special relativity is that, unlike Einstein's later theory of general relativity—which essentially deals with gravity—there's nothing remotely controversial about it. No professional physicists, including those pursuers of a "theory of everything" who dislike general relativity, would ever dispute the validity of special relativity. If there's any argument around it at all (among people who understand it correctly, that is) then it's the extent to which it really was Einstein's discovery, or whether other physicists got there before him.

For non-physicists, on the other hand—which of course includes the great majority of writers and readers of science fiction—special relativity is fraught

Fig. 4.8 Albert Einstein was responsible for many things, but—when you examine his theory closely—limiting the rate at which space travellers can get from A to B wasn't one of them (public domain image)

with problems, simply because it's so counter-intuitive. The central difficulty is implicit in the very word "relativity", because in Einstein's view of the universe, it's not just speed that's relative but the way time itself passes. The rate at which a clock ticks depends on how fast it's moving relative to the person observing it.

Most SF, whatever its intended audience, simply ignores the effects of relativity altogether. In the case of movies aimed at the mass market, this is for the obvious reason that most viewers wouldn't be able to understand it, let alone accept it. But even in serious novels aimed at more scientifically savvy readers, attempting to include all the ramifications of relativity is likely to get in the way of whatever story the author is trying to tell. The only exception is when that story itself revolves around the effects of relativity, as in Robert A. Heinlein's *Time for the Stars* (1956) or Joe Haldeman's *The Forever War* (1974). Even then, authors are as likely as not to get the fine details wrong. As reviewer James Davis Nicoll wrote of Heinlein's book:

> One must admit that he did not grasp relativity. When he goes into any detail on the matter, it becomes really, really clear that he didn't understand [39].

Given the counter-intuitive nature of relativity, this is entirely understandable. The same can't be said of other scientific errors perpetrated by SF authors—since, for example, the basic physics of orbital motion is pure high-school stuff, and there's really no excuse (other than laziness) for getting it wrong. But few students encounter relativity before their first year at university, and even then it's often the hardest subject to get their heads around. Ultimately, you just have to accept that it's the way the universe works, even if it's not the way you feel the universe *ought to* work.

To return to our starting point, the basic problem revolves around the speed of light—and it's worth thinking through just what this means. It's known that the photons making up a beam of light have zero mass, which means they have zero inertia—or zero resistance to any force that's applied to them.

Naively, this suggests that if you push on a photon, even with an infinitesimally tiny force, it would instantly fly off at infinitely high speed. That's basically what does happen, except that the speed isn't infinite. It's a very specific finite speed, around 300,000 km/s, which is generally known as the "speed of light", or c. We just have to accept that this completely arbitrary velocity is what stands in for "infinite speed" in our universe. Any object, not just a photon of light, that has zero mass must necessarily travel at this speed all the time. It's a point that Isaac Asimov makes in his 1967 short story "The Billiard Ball":

> A completely massless object can move in only one way … motion at the speed of light. Any massless object, such as … a photon, must travel at the speed of light as long as it exists. In fact, light moves at that speed only because it is made up of photons [40].

That last point is subtle but important. It's not that the "cosmic speed limit" is the speed of light, but that light travels at the speed it does *because* it's the speed limit. But why does the speed limit have that particular value, c, rather than being equal to infinity as you might expect?

Let's think this through. The definition of "speed" is the distance travelled divided by the time taken. So, by this definition, an infinite speed would mean that you reached your destination—whatever its distance—in zero time. But it turns out this is exactly what *does* happen in the case of light, or anything travelling at speed c. It's the single most important fact—albeit a totally non-intuitive one—lying at the heart of Einstein's theory. Subjectively, you can travel at any speed you like, in the sense of getting from point A to point B in as short a time as you care to imagine. Einstein doesn't place any limits on this at all. But if you *were* to travel from one star system to another in the space of a heartbeat, then an observer who remained stationary at your starting point would measure you as having travelled at the finite speed c.

This sounds crazy—which is why so many people choose to dispute or ignore relativity—but it's the way our universe happens to be constructed. The rate at which time passes is different for different observers moving at different speeds relative to each other. This has obvious consequences for high-speed interstellar travel, which we'll come to in a moment, but it also has a practical impact even on long-distance space missions today. This is because light, from any point of view but its own, is constrained to travel at that particular speed c. The same is true of any other form of electromagnetic radiation, such as a radio wave, which leads to the now familiar phenomenon of the "long distance communications delay".

It's an effect that was much less familiar, to most people at least, back in 1962, when Isaac Asimov made it the central plot driver of his short story "My Son, the Physicist". Here is how one of the story's characters encapsulates the problem:

> At the present moment Pluto is just under 4 billion miles away. It takes 6 hours for radio waves, travelling at the speed of light, to reach from here to there. If we say something, we must wait 12 hours for an answer [41].

More than half a century later, an almost identical situation occurred in the real world, when NASA's New Horizons probe flew past Pluto in 2015 (see Fig. 4.9). This actually took place closer to Earth than Asimov assumed in his story—not because he was wrong, but because Pluto moves in an eccentric orbit so it's not always at the same distance from us. In the case of the New Horizons encounter, the one-way time delay between the probe sending data back and it being received on Earth was 4.5 h.

Communication delays aside, the finite speed of electromagnetic waves does have its practical uses. It's the reason that a radar system can measure the precise distance to a target, based on the time taken for a signal to reach the target and then bounce back to the radar dish. Conventionally this uses radio waves, but the same principle can be applied to any other part of the electromagnetic spectrum. In *Project Hail Mary*, for example, Andy Weir's protagonist puts together an improvised detector system based on infrared radiation—and then uses it to estimate the range to another spacecraft:

> I stare at the screen with the stopwatch ticking away in my hand. Soon, I see the blip again. 28 seconds.... It's 14 light-seconds away (14 seconds to get there, 14 seconds to get back equals 28 seconds). That works out to about 4 million kilometres [42].

Fig. 4.9 Artist's impression of NASA's New Horizons probe flying pas the dwarf planet Pluto (NASA image)

The mathematics here is all comfortingly familiar. When you send out a beam of photons travelling at speed *c* kilometres per second, they take time *d*/*c* seconds to reach a destination *d* kilometres away. There's nothing counter-intuitive there. It's only when we turn to *people*—in the form of spaceship crews—travelling at speeds close to *c* that things start to get complicated.

Actually, many accounts of relativity—occasionally in fiction as well as in popular science books—make it sound even more complicated than it needs to be. This comes from dwelling too much on what a "stationary" observer, left behind at the point of departure, would see if they looked at a rapidly departing spaceship through a telescope. But that's really not relevant to what the space travellers themselves experience, so we will do better to ignore it altogether. This leaves us with a simpler situation, which—although still far from obvious—is not too difficult to grasp.

To start with, what the crew *themselves* experience as they fly at high speed from one star system to another is virtually the same as you would expect if you knew nothing at all about relativity. The more thrust they apply to their spacecraft, the faster it goes, and the more rapidly they approach their destination. Depending on how powerful their space drive is, the can make the journey time as short as they like (although, for practical reasons, it would be uncomfortable to accelerate at much more than 1 g for a sustained period of time).

To pick a concrete example, the spaceship in Andy Weir's *Project Hail Mary* is designed to travel the 12 light-years from the Solar System to Tau Ceti at a constant acceleration of 1.5 g. The journey takes just under 4 years, much less than the 12 years that a strict "one light-year per year" speed limit might suggest. Although the novel is science fiction, the travel time isn't. It comes straight from Einstein's equations of relativity.

So that's the big myth of relativity debunked; it doesn't limit the rate at which you can travel from A to B. However, it does come with a couple of "catches" that can confuse the situation when you first encounter them. To start with, although it appears the ship has travelled "faster than the *speed* of light", it hasn't actually travelled "faster than light" itself. Light that left Earth at the same time as the ship would still reach Tau Ceti before it. In other words, if the crew looked back at Earth through a telescope, they would see it as it looked *after* they left, not before.

The other catch—and it's the biggest one of all—is that during the 4 years the ship takes to make the trip, something like 13 years would have passed on Earth. There's no point trying to justify this using logic or common sense, because it can't be done. It's just the way the universe is, and we have to accept it. Time runs at different rates for observers moving at different speeds.

As interstellar travel times go, the 4 years experienced by the human-crewed ship in *Project Hail Mary* sounds pretty impressive. But it's actually longer than it needs to be, because it's limited by the maximum acceleration that humans can tolerate. Without a crew, a spacecraft could make the journey in an even shorter time—as is the case for the miniature robotic probes that feature in the same novel:

> They accelerate at 500 g's until they reach a cruising speed of 0.93 *c*. It'll take over 12 years to get back to Earth [from Earth's point of view], but all told the little guys will only experience about 20 months [43].

Even at a human-tolerable acceleration, huge distances could be covered in a manageable length of time if the thrust is kept up for long enough. In principle (though probably not in practice, as we saw earlier), this could be done with a Bussard ramjet—such as the one featured in Larry Niven's novel *A World out of Time*. Here, the protagonist muses about reaching the centre of the galaxy, around 26,000 light-years away:

> The relativity equations work better for me the faster I go.… It works out that I can reach the galactic hub in 21 years, ship's time, if I hold myself down to one gravity acceleration [44].

Of course, you expect writers like Larry Niven and Andy Weir to get their scientific facts right. More surprising, however, is the case of L. Ron Hubbard—who is best known in hindsight (or maybe "notorious" is a better word) as the founder of the Church of Scientology. Before this, however, he was a fairly prolific writer of science fiction, and in one novel he managed to get the facts of relativity at least approximately correct. Called *Return to Tomorrow* and dating from 1950, its storyline is based so heavily on the idea of relativistic "time dilation" that Hubbard actually quotes the mathematical equation for it. This is shown in Fig. 4.10 in the form it took when the novel was serialized (under the alternative title of *To the Stars*) in *Astounding Science Fiction* magazine [45].

This is the basic Einstein-Lorentz formula for the time T_v measured on a spaceship travelling at speed v relative to its point of origin. It's equal to the time T_0 measured by a stationary observer at the origin, multiplied by the square root of 1 minus the square of the ratio of v to c, the speed of light. This square root is always less than 1, so T_v is always less than T_0.

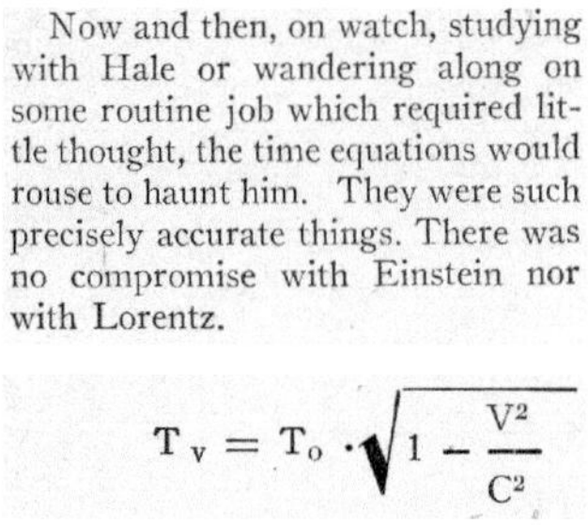

Now and then, on watch, studying with Hale or wandering along on some routine job which required little thought, the time equations would rouse to haunt him. They were such precisely accurate things. There was no compromise with Einstein nor with Lorentz.

$$T_V = T_0 \cdot \sqrt{1 - \frac{V^2}{C^2}}$$

Fig. 4.10 Extract from a story by L. Ron Hubbard that appeared in the February 1950 issue of *Astounding Science Fiction*, featuring the (mathematically correct) equation for relativistic time dilation (Internet Archive)

References

1. J.W. Campbell, Not Quite Rockets, in *Astounding Science Fiction*, (April 1944), p. 99
2. R.A. Heinlein, *The Man Who Sold the Moon* (Pan Books, London, 1963), p. 141
3. A.C. Clarke, *The Exploration of Space* (Kindle Edition), loc. 434
4. J. Pournelle, *A Step Farther Out* (Star Books, London, 1981) pp. 47-9
5. S. Baxter, *Voyage* (Harper Collins, London, 2015) p. 8
6. C. Hadfield, *The Apollo Murders* (Kindle Edition), loc. 1990
7. C. Lee, The Test Stand, in *Astounding Science Fiction*, (March 1955), p. 67
8. C. Hadfield, *The Apollo Murders* (Kindle Edition), loc. 2027
9. C. Hadfield, *The Apollo Murders* (Kindle Edition), loc. 569
10. I. Asimov, *The Martian Way* (Panther Books, London, 1973), pp. 13–14
11. Bergita and Urs Ganse, *The Spacefarer's Handbook* (Springer, Berlin, 2020), pp. 35–36
12. A.C. Clarke, *2001: A Space Odyssey* (Arrow Books, London, 1968), p. 49
13. A.C. Clarke, *Prelude to Space* (New English Library, London, 1968), pp. 19–20
14. E. Ralph, SpaceX CEO Elon Musk Details Orbital Refuelling Plans for Starship Moon Lander, https://www.teslarati.com/spacex-elon-musk-starship-orbital-refueling-details/
15. A. May, *The Space Business* (Icon Books, London, 2021), pp. 74–75
16. A.C. Clarke, *Islands in the Sky* (Kindle Edition), loc. 1758
17. B. Clegg, *Final Frontier* (St. Martin's Press, New York, 2014), pp. 59–60
18. A.C. Clarke, *The Fountains of Paradise* (Del Rey Books, New York, 1980), p. 55
19. A.C. Clarke, *The Fountains of Paradise* (Del Rey Books, New York, 1980), p. 300
20. R.L. Forward, Beanstalks, in *Indistinguishable from Magic*, (Baen Books, New York, 1995), pp. 61–62
21. C. Hadfield, *The Apollo Murders* (Kindle Edition), loc. 577 & 3779

22. Z. Webb-Mack, A Brief History of Ion Propulsion, https://solarsystem.nasa.gov/news/723/a-brief-history-of-ion-propulsion/
23. A. Weir, *The Martian* (Del Rey, New York, 2014) p. 2
24. A. Weir, *The Martian* (Del Rey, New York, 2014) p. 278
25. L. Zyga, Plasma Rocket Could Travel to Mars in 39 Days, https://phys.org/news/2009-10-plasma-rocket-mars-days.html
26. A.C. Clarke, *Prelude to Space* (New English Library, London, 1968), p. 67
27. G. Benford, *The Martian Race* (Orbit Books, London, 2000), pp. 70–72
28. S. Baxter, *Voyage* (Harper Collins, London, 2015), pp. 24–25
29. D. Graham-Rowe, Nuclear Fusion Could Power NASA Spacecraft, https://www.newscientist.com/article/dn3294-nuclear-fusion-could-power-nasa-spacecraft/
30. J. Pournelle, *A Step Farther Out* (Star Books, London, 1981) p. 58
31. A.C. Clarke, *Imperial Earth* (Pan Books, London, 1978), p. 103
32. L. Niven, World of Ptavvs, in *Worlds of Tomorrow*, (March 1965), pp. 52–53
33. L. Niven, "Bigger than Worlds", in *A Hole in Space* (Orbit Books, London, 1984), p. 113
34. L. Niven, *A World out of Time* (Orbit Books, London, 1988) p. 9
35. J. Ouellette, Study: 1960 Ramjet Design for Interstellar Travel Is Unfeasible, https://arstechnica.com/science/2022/01/study-1960-ramjet-design-for-interstellar-travel-a-sci-fi-staple-is-unfeasible/
36. A.C. Clarke, *The Wind from the Sun* (Pan Books, London, 1983), p. 50
37. R.L. Forward, Fading into Blackness, in *Indistinguishable from Magic*, (Baen Books, New York, 1995), pp. 136–137
38. Starshot, *Breakthrough Initiatives*, https://breakthroughinitiatives.org/initiative/3
39. J.D. Nicoll, SF Novels That Get Special Relativity All Wrong, https://www.tor.com/2018/12/05/sf-novels-that-get-special-relativity-all-wrong/
40. I. Asimov, The Billiard Ball, in *Asimov's Mysteries*, (Panther Books, London, 1971) pp. 249-50
41. I. Asimov, My Son the Physicist, in *Nightfall Two*, (Panther Books, London, 1971) p. 179
42. A. Weir, *Project Hail Mary* (Kindle Edition), loc. 7194
43. A. Weir, *Project Hail Mary* (Kindle Edition), loc. 4872
44. L. Niven, *A World out of Time* (Orbit Books, London, 1988), p. 32
45. L. Ron Hubbard, To the Stars, Part 1, in *Astounding Science Fiction*, (February 1950), p. 42

5

Living in a Vacuum

From a terrestrial perspective, the space environment can seem disconcertingly alien in many different ways, even though the basic laws of physics are the same. The primary focus of this chapter is on how physics works outside the Earth's atmosphere—from the aerodynamics, or lack of it, of space vehicles and the design of life support systems to the extreme cold and potentially deadly radiation of interplanetary space. Also addressed is the question of how scientists—or space travellers—can use suitably designed instruments to study objects that may be millions of kilometres, or even many light-years, away. As in previous chapters, the physical principles are brought to life with the aid of quotations from the science fiction of Arthur C. Clarke, Andy Weir and others.

5.1 Outside the Atmosphere

One of the things we take most for granted about the Earth is its atmosphere. This is a relatively thin layer of air—or, to put it more scientifically, a mixture of gases—that's held close to the planet's surface by gravity. As human beings, the most important consequence of the atmosphere is that it allows us to breathe, and we'll come back to this aspect a little later. First, however—because this is a book primarily about space travel—it's worth taking a look at the atmosphere in that context.

We saw in the previous chapter how jet engines, unlike rockets, rely on air sucked in from the atmosphere in order to work. An aircraft also needs the atmosphere for another reason too; as it travels along and air flows over its

A. May, *How Space Physics Really Works*, Science and Fiction,
https://doi.org/10.1007/978-3-031-33950-9_5

wings, this generates the lift needed for it to stay airborne. When an aircraft banks in order to make a turn, it's not because it looks cool (as the directors of outer space movies like *Star Wars* seem to think), but because the aerodynamic lift forces act perpendicular to the wings. In order to turn, it helps if those forces act in the direction you want to go, rather than vertically upwards.

Even before the days of aircraft, people had an intuitive understanding of this kind of aerodynamic manoeuvring, from watching birds in flight. So it's deeply ingrained in our picture of how the world works—and something we need to forget all about if we're going to consider the case of vehicles travelling in the vacuum of outer space.

One major difference concerns the process of changing direction—which, as we just saw, is usually achieved within the Earth's atmosphere by the act of banking. But this just won't work in outer space, where aerodynamic forces like lift simply don't exist. Instead, a spacecraft can only change direction by applying rocket thrust—and, whatever the *Star Wars* people think, it doesn't have to bank on its side to do this.

Another consequence of the lack of air in outer space—and, again, one that's not always appreciated in mass market science fiction—is that there's no need for a spacecraft to be streamlined in the same way that an aircraft is. This is a point that Arthur C. Clarke made in his novel *The Sands of Mars*, from 1951:

> Gibson had never become reconciled to the loss of the sleek, streamlined spaceships which had been the dream of the early 20th century. The glittering dumbbell hanging against the stars was not *his* idea of a space liner .... Of course, he knew the familiar argument—there was no need for streamlining in a ship that never entered an atmosphere, and therefore the design was dictated purely by structural and power plant considerations [1].

When Clarke referred to streamlined spaceships as being "the dream of the early 20th century", he was probably thinking of the innumerable SF magazine covers of the 1930s and 40 s, which invariably portrayed just such ships. However, by the time he wrote *The Sands of Mars*, depictions of this type were becoming passé, at least in the case of more upmarket SF magazines such as *Galaxy*. A far more realistic example dating from 1953—still well before the first real-world spacecraft was launched—is shown in Fig. 5.1.

If the Earth has an atmosphere and space doesn't, where does the boundary between the two lie? It's not an easy question to answer, because there's no sudden change from one to the other. As you ascend in altitude above the Earth, the air gradually gets thinner and thinner, but there are still a few

Fig. 5.1 The cover of *Galaxy Science Fiction* for September 1953 shows a realistically *un*-streamlined spacecraft, as well as realistic-looking spacesuits on the astronauts (public domain image)

molecules of gas to be found even in what may look like a vacuum for all practical purposes. As Larry Niven puts it in his 1965 short story "Wrong Way Street":

> The Earth's atmosphere goes way past the Moon… of course, it gets pretty thin. The idea is that the Earth's atmosphere ends where its density drops to the density of the surrounding space [2].

If you're looking for an objective definition of the Earth's atmosphere, the one given by Niven is as good as any. But it's purely academic, because the "density of the surrounding space" that he refers to is extremely low. It's true that the space between planets does contain a few particles of matter, but only about 5 of them per cubic centimetre. That's basically nothing, when compared to the 26 quintillion particles in the same volume of air at sea level on the Earth—or even the 10 billion particles per cc in a typical "vacuum" chamber in a professional laboratory.

As we saw when we discussed the concept of Bussard ramjets in the previous chapter, there are even a few atoms in the space between stars. This is even more tenuously populated than the space between planets in the Solar System, with typical interstellar densities being just a few atoms per cubic *metre* rather per cubic centimetre. As infinitesimally low as this sounds, there are circumstances under which it means that something we dismissed as unnecessary a moment ago—the streamlining of a spaceship—might actually be a good idea after all.

The key lies in just how fast the ship is travelling. We know from the first chapter of this book that kinetic energy is proportional both to mass and to the square of velocity. Any interstellar atoms that a ship might smash into will have a very small mass, but their relative velocity may be very high—close to the speed of light. Remembering that we then have to square this velocity to get the energy the atoms transfer to the ship, it's perfectly possible that they could give it quite a battering. So, for a very fast-moving spaceship, maybe "aerodynamic" streamlining is a useful thing to have after all.

It's a point that Andy Weir, at least, thought was worth incorporating into the design of the eponymous interstellar ship in *Project Hail Mary* (2021):

> The *Hail Mary* has always looked like something out of a Heinlein novel. Shiny silver, smooth hull, sharp nose cone. Why do all that for a ship that'll never have to deal with an atmosphere? Because of the interstellar medium. There's a teeny, tiny amount of hydrogen and helium wandering around out there in space. It's on the order of one atom per cubic centimetre, but when you're traveling near the speed of light, that adds up [3].

The small but finite density of the interstellar medium is what makes a Bussard ramjet possible in principle, but it may also set a practical limit on the speed it can travel at. This is a subject that Isaac Asimov raised in his non-fiction book *The Relativity of Wrong*, from 1988:

> It may not be practical for an interstellar spaceship to go faster than a fifth the speed of light. After all, the faster we go, the more difficult it is to avoid collisions with small objects and the more damage such collision will wreak. Even if we are fortunate enough to miss all sizable objects, we can scarcely expect to miss the dust and individual atoms that are scattered throughout space. At two-tenths the speed of light, dust and atoms might not do significant damage even in a voyage of 40 years, but the faster you go, the worse it is—space begins to become abrasive [4].

Coming back down to Earth, or nearly so, a more familiar example of a spacecraft experiencing abrasion is the re-entry of a capsule into the Earth's atmosphere. The main effect here is a thermal one, as former astronaut Chris Hadfield describes in his 2021 novel *The Apollo Murders*:

> *Pursuit* was 75 miles high, just beginning to touch the outer atmosphere .... On the outside of the capsule, bad things were starting to happen. *Pursuit* was slamming into the rarefied air molecules with so much energy that it was ripping electrons free and tearing at its belly shield, burning off the outer layer. The mix of ionized gases clung to *Pursuit*'s skin in a sheath of blowtorch flames; an electric plasma field, glowing yellow and orange and red, enveloped the ship in a hypersonic fireball [5].

Once again, the biggest problem here isn't the density of material the capsule encounters, but its speed relative to the atmosphere that it's entering. This isn't only an issue for a spacecraft returning to Earth, but also for one that's entering the atmosphere of a different planet—even the much more tenuous atmosphere of Mars (see Fig. 5.2).

5.2 Life Support

For inhabitants of the Earth like ourselves, the most important fact about the atmosphere is that it keeps us alive. It's composed of a mixture of gases, containing around 21% oxygen and 78% nitrogen, with the remaining 1% made up of various other elements. However, from our point of view, it's the atmosphere's oxygen content that really matters.

Without diving down into the details of human biochemistry, it's a basic fact that all terrestrial animals like ourselves need oxygen to breathe. That's simple enough to say, but not always so easy in practice, because oxygen is a highly reactive gas. This means that it's keen to combine with other elements—to produce oxides and other oxygen-containing compounds—and if left to itself it would very quickly disappear from the atmosphere. Fortunately for us, the Earth's atmospheric oxygen is constantly being replenished by plants and bacteria, which absorb water (H_2O) and carbon dioxide (CO_2), extract the hydrogen and carbon for their own use, and then "exhale" the oxygen.

So how are space travellers going to maintain an adequate supply of oxygen inside a spaceship? One way would be to copy the way that it's done on Earth, by cultivating special plants for the purpose. This is the approach used on the

Fig. 5.2 Artist's impression of a space probe entering the rarefied upper atmosphere of Mars (NASA image)

"Venus Equilateral" space station (mentioned previously in the chapter on gravity), as envisioned by SF author George O. Smith in his short story "QRM Interplanetary", from 1942:

> What better purifying machine is there than a plot of grass? … We breathe oxygen, exhale CO_2. Plants inhale CO_2 and exude oxygen. An air plant means just that. It is a specialized type of Martian sawgrass that is more efficient than anything else in the system for inhaling dead air and revitalizing it [6].

In the real world, for example on board the International Space Station (ISS), a rather less exotic method is used to extract oxygen from water, via an electrically driven process called electrolysis. Even more simply, for short duration missions it's possible to carry along all the oxygen the astronauts are going to need in pressurized tanks. This is what was done on the Apollo missions to the Moon, where in fact the astronauts breathed pure oxygen at reduced pressure, rather than the oxygen-nitrogen mix of the Earth's atmosphere. This has the benefit of simplicity, since the human body has no need for nitrogen, but it does mean a person can't just switch instantly from one type of atmosphere to another. Here's how Chris Hadfield describes the transition in *The Apollo Murders*:

> Over the next few hours, the nitrogen in his blood would be gradually replaced by oxygen; that way, when he got to orbit and popped his helmet off in the low pressure all-oxygen atmosphere, his blood nitrogen wouldn't suddenly bubble and give him the bends [7].

In fact, it's still true of extravehicular activity (EVA) spacesuits to this day that astronauts wearing them breathe pure oxygen, even if they're working outside the ISS with its Earth-like oxygen-nitrogen atmosphere. This is something that Andy Weir refers to in *Project Hail Mary*:

> Internal pressure is 400 hectopascals—about 40% of Earth's atmosphere at sea level. That's normal for spacesuits.... On space stations back home, astronauts have to spend hours in an airlock slowly acclimating to the low pressure needed for the EVA suit before they can go out ... space stations around Earth have a full atmosphere of pressure in case the astronauts have to abort and return to Earth in a hurry [8].

Of course, there's a lot more to maintaining a healthy environment on the ISS than having enough oxygen to breathe. Harmful or unpleasant gases have to be removed from the atmosphere, as well as recycling as much water as possible to avoid it going to waste. All of this is done by a system called the Environmental Control and Life Support System, or ECLSS (see Fig. 5.3).

Right at the start of this book, we listed a number of scientific misconceptions commonly encountered in movies and TV shows, courtesy of the *TV Tropes* website. One of these misconceptions was that "after a hole is opened in a space ship's outer structure, it has about the same effects as a nearby tornado". That's not the way it would happen at all, particularly if the hole was very small—as was the case with the 2 mm hole discovered in a Soyuz capsule that was docked to the ISS in 2018. The first indication of the hole came when ground controllers noticed a drop in the station's air pressure, but the rate of air loss was so gradual they didn't even wake the crew. The next day, the astronauts located the hole and easily fixed it with an epoxy sealant [9].

The Soyuz hole was probably the result of shoddy workmanship, but it might just as well have been caused by a micrometeoroid impact. This is just what does happen to a spaceship travelling between the Earth and Mars in Arthur C. Clarke's novel *The Sands of Mars*. The crew notice it because, as in the real-world ISS case, it causes a slight loss of air pressure. Again, as in the

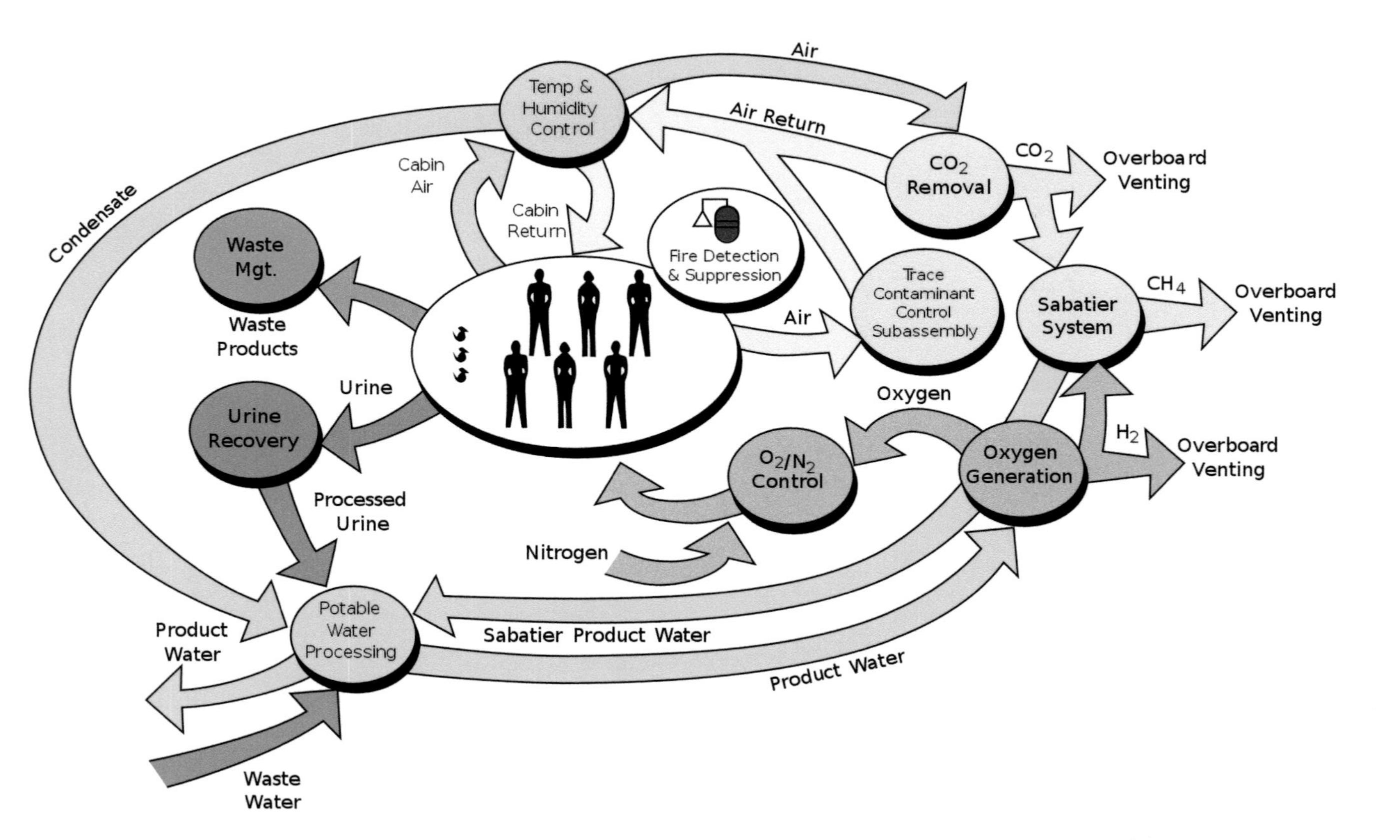

Fig. 5.3 A flow diagram showing the various components of the ECLSS system on board the ISS (NASA image)

real world, they're not overly alarmed by it, because "punctures due to meteoric dust happened 2 or 3 times a year on a ship of this size". Awkwardly for them, however, the hole happens to be inside the cabin of their VIP guest—a writer who they're hoping will give them positive publicity—and the last thing they want is for him to tell the world his cabin wall was holed by a meteoroid. Fortunately, they're able to repair the leak without him ever becoming aware of it:

> We shut off the ventilators and found the leak right away with the good old smoky candle technique. A blind rivet and a spot of quick drying paint did the rest [10].

So, as so often, Arthur C. Clarke was getting his scientific facts right long before many of the Hollywood directors who consistently get them wrong were even born. Another author who can separate fact from fiction is Andy Weir, whose protagonist actually *wants* to let the air out of his spaceship at one point in *Project Hail Mary*—and ends up being frustrated by the time it takes:

> It takes a surprisingly long time for a ship to lose all its air. In movies, if there's a little breach everyone dies immediately.... But in real life, air just doesn't move that fast. The emergency valve on the airlock is 4 centimetres across. Seems like a pretty big hole to have in your spaceship, right? It took 20 minutes for the ship's air pressure to drop to 10% of its original value. And it's dropping very slowly now [11].

There's another aspect to atmospheric life support that's less obvious than the need for fresh oxygen. While humans inhale oxygen, they also exhale CO_2—and too much of the latter is just as fatal as too little of the former. The natural amount of CO_2 in the Earth's atmosphere is very small—only about 400 parts per million, or 0.04 per cent. On the other hand, as much as 4 per cent of the air that we breathe out is CO_2—about 100 times the normal ambient concentration. If nothing is done to get rid of this, it will steadily build up in the atmosphere until—at a concentration of around 7%—it becomes fatal.

Obviously steps need to be taken to remove excess CO_2 from the enclosed atmosphere of a spacecraft, in a process generically referred to as "CO_2 scrubbing". Even Jules Verne was aware of this necessity back in 1870, when he wrote his SF masterpiece *Around the Moon*. Here is how he describes the solution that his space travellers arrive at:

> It was not enough to renew the oxygen expended; it was necessary to absorb the carbon dioxide produced by exhalation. For a dozen hours now, the atmosphere of the projectile had charged itself with this deleterious gas.... But Captain Nicholl hastened to remedy this state of affairs. He placed on the bottom of the projectile several vessels containing potassium hydroxide, which he stirred for a time, and this substance, greedy for carbon dioxide, absorbed it completely and so purified the air [12].

The process described by Verne is a perfectly valid one—either using potassium hydroxide as he suggests, or a related chemical with a similar propensity to lock onto any free CO_2 molecules that happen to be around. By snatching them out of the atmosphere, such chemicals act as the simplest form of CO_2 scrubber—and one of them, lithium hydroxide, was used on the Apollo missions. The ISS uses a more sophisticated mechanical system, but simple chemical scrubbers are still employed in spacesuits. It's a point that Andy Weir makes in *The Martian* (2014):

> After a while, the CO_2 absorbers in the suit were expended. That's really the limiting factor to life support. Not the amount of oxygen you bring with you, but the amount of CO_2 you can remove. In the hab, I have the oxygenator, a large piece of equipment that breaks apart CO_2 to give the oxygen back. But the spacesuits have to be portable, so they use a simple chemical absorption process with expendable filters [13].

In contemporary fiction, authors can refer to spacesuits and surface habitats—such as the "hab" that Weir mentions—without having to tell the reader what they are and why they're necessary. We all know *that* much about the space environment, at any rate. But in the past it was different, and in Robert A. Heinlein's short story "Misfit", from 1939, he had to go into considerable detail about both subjects. The story focuses on an expedition to a moderately large asteroid, around 150 kilometres in diameter, and here is how one of the characters describes the spacesuits that the crew members have to wear:

> "This is a standard service type, general issue, Mark IV, Modification 2." He grasped the suit by the shoulders, and shook it out so that it hung like a suit of long winter underwear with the helmet lolling helplessly between the shoulders of the garment. "It's self- sustaining for 8 hours, having oxygen supply for that period. It also has a nitrogen trim tank and a carbon-dioxide-water-vapour cartridge filter.... The suit is woven from glass fibre laminated with non-volatile asbestocellutite. The resulting fabric is flexible, very durable; and will turn all rays normal to solar space outside the orbit of Mercury. It is worn over your

> normal clothing, but notice the wire-braced accordion pleats at the major joints. They are so designed as to keep the internal volume of the suit nearly constant when the arms or legs are bent. Otherwise the gas pressure inside would tend to keep the suit blown up in an erect position, and movement while wearing the suit would be very fatiguing."

Heinlein's explorers go on to create a habitat on the asteroid, combining natural features with artificially constructed ones:

> The captain selected a little bowl-shaped depression in the hills, some thousand feet long and half as broad, in which to establish a permanent camp. This was to be roofed over, sealed, and an atmosphere provided. In the hill between the ship and the valley, quarters were to be excavated: dormitories, mess hall, officers' quarters, sick bay, recreation room, offices, store rooms and so forth [14].

This is reminiscent of the "domes" that became ubiquitous in later SF, including Arthur C. Clarke's *The Sands of Mars* that we've already encountered a couple of times in this chapter. Although real-world Mars habitats still lie in the future, there's a good chance they will indeed take the form of pressurized domes (see Fig. 5.4).

5.3 Vacuum Physics

There's one bit of vacuum physics that everyone knows, thanks to the tagline of Ridley Scott 1979 movie *Alien*: "In space, no one can hear you scream". In other words, sound doesn't travel through the vacuum of outer space. It's a phenomenon that the protagonist of Arthur C. Clarke's short story "The Other Side of the Sky" (1957) experiences first hand, when he's briefly exposed to a vacuum without the benefit of a spacesuit:

> The only thing I can be certain of is that uncanny silence. It is never completely quiet in a space station, for there is always the sound of machinery or air pumps. But this was the absolute silence of the empty void, where there is no trace of air to carry sound [15].

Sound, like so many other physical effects that occur within the Earth's atmosphere, is something we all tend to take for granted. But it simply couldn't exist without a medium like air to travel through. In another of Clarke's short stories, "Silence Please" from 1954, he has one of the characters explain the basic physics of sound in the following way:

Fig. 5.4 Future habitats on Mars may well take the form of domes, as prophesied by numerous SF stories (NASA image)

> I do not know if you have ever considered the nature of sound. Suffice to say that it consists of a series of waves moving through the air. Not, however, waves like those on the surface of the sea … those waves are up and down movements. Sound waves consist of alternate compressions and rarefactions [16].

Without any air pressure to be varied in this way, there's no way for a sound wave to propagate through a vacuum. Another atmospheric effect, which we similarly take for granted, can prove very confusing for people trying to learn basic physics. We saw in an earlier chapter how all objects are subject to the same downward acceleration due to gravity. This implies that all objects should fall at the same rate, and yet we know that they don't. Flakes of dust, for example, tend to fall very slowly compared to larger objects. This, however, isn't because they experience a weaker pull of gravity, but because there's a compensating upward force on them due to the air itself.

This is no longer a consideration on the Moon, where there's no atmosphere to speak of. There, although objects fall more slowly than on Earth due to the weaker lunar gravity, they do all fall at the same rate. It's a point that Arthur C. Clarke made in his 1955 novel *Earthlight*:

> Most objects fell too slowly here in this low gravity for anyone accustomed to conditions on Earth. But dust fell much *too* quickly—at the same rate as anything else, in fact—for there was no air to check its descent [17].

Another consequence of the Moon's lack of an atmosphere is so obvious, even to non-physicists, that it's caused a lot of confusion in some quarters. If there's no air on the Moon, there can't be any wind, and everyone knows that the wind is what makes a flag fly. Yet the flag planted on the Moon by the crew of Apollo 11 in 1969 really does appear to "fly", rather than hanging straight down as it ought to in a vacuum. To some people, this is clear evidence that the event was actually filmed here on Earth.

Things, however, aren't quite as they seem. The people who planned the Apollo missions weren't stupid, and they anticipated the problem of an unimpressively limp flag in advance. So they took the precaution of stiffening its upper edge with a horizontal rod (see Fig. 5.5).

The trick wasn't even original to NASA. It had already been used in an SF context by Robert Heinlein in his novel *Rocketship Galileo*, as long ago as 1947:

> On a short and slender staff the banner of the United Nations and the flag of the United States whipped to the top. No breeze disturbed them in the airless waste—but Ross had taken the forethought to stiffen the upper edges of each with wire [18].

It's also no surprise that Arthur C. Clarke, as ever a stickler for scientific accuracy, used a similar idea in his 1961 novel *A Fall of Moondust*:

> The illusion was excellent, for the lines of pennants draped around the embarkation building were stirring and fluttering in a non-existent breeze. It was all done by springs and electric motors, and would be very confusing to the viewers back on Earth [19].

Fig. 5.5 The flag planted by the crew of Apollo 11 appears to fly in a non-existent lunar wind, but this is an illusion created by a clever engineering design (NASA image)

Another point that Clarke makes in the same novel concerns the temperature that water boils at. We often think of this as a hard-and-fast 100 C, but it's only this temperature if the air pressure has the value that it does at ground level on Earth. If the pressure is lower, for example in an artificial habitat on the Moon, then water will boil at a lower temperature. So raising the pressure to terrestrial levels has at least one major advantage, as Clarke points out:

> Now that he had put up the cabin pressure, water must be boiling at nearly its normal, sea-level temperature back on Earth. At least they could have some hot drinks—not the usual tepid ones [20].

In the extreme case of a vacuum, water will boil spontaneously even at room temperature. This is a piece of scientific knowledge that Larry Niven's protagonist puts to good use in the novel *A World out of Time* (1976). Safely encased in a spacesuit, he's checking out a spaceship on the Moon that may or may not be properly pressurized: "The ship's gauges said air. The suit's said vacuum. Which was right?" His solution is simply to turn on a water tap; since the water doesn't boil, he knows the ship must contain air [21].

Like air pressure, the temperature we normally talk about, for example in the context of the weather, is a property of the Earth's atmosphere; it's a measure of how fast the air molecules are moving. But there's a temperature in the vacuum of space, too, due to infrared radiation emanating from the Sun or other bodies in the Solar System. Just what this temperature is forms the subject of a debate in Jules Verne's novel *Around the Moon*.

Verne's space travellers discuss two rival theories that were current at the time the novel first came out, in 1870. One scientist, Joseph Fourier, estimated the temperature of space to be -60 C, while another, Claude Pouillet, considered it to be a much colder -160 C. To resolve the issue, Verne's crew measure the temperature for themselves, using a specially designed thermometer they protrude from their capsule when it's out of direct sunlight behind the Moon. The measurement they obtain tends to support Pouillet's theory, showing a value of -140 C. In hindsight, this isn't too wide of the mark; according to Wikipedia, the temperature on the night side of the Moon is around -170 C [22].

The fact that space is so cold means that heat leaks away very quickly. This is basically a consequence of the second law of thermodynamics, which states that heat always flows from warmer bodies to cooler ones—also known as the principle of entropy. It's something that Mark Watney, protagonist of Andy Weir's novel *The Martian*, comes up against when he tries to remain in an unheated rover in the intensely cold environment of Mars:

> I was fine for a while. My own body heat plus three layers of clothing kept me warm, and the rover's insulation is top-notch.... But there's no such thing as perfect insulation, and eventually the heat left to the great outdoors, while I got colder and colder.... All my brilliant plans foiled by thermodynamics. Damn you, entropy! [23]

At the other extreme, if the insulation is too good, and there's an internal source of heat, the temperature can rise to uncomfortable levels even if it's freezing cold outside. This is what happens in Arthur C. Clarke's *A Fall of Moondust*, when the occupants of a vehicle find themselves buried under a huge quantity of the titular dust:

> We're blanketed with this stuff, and it's about the best insulator you can have. On the surface, the heat our machines and bodies generated could escape into space, but down here it's trapped. That means we'll get hotter and hotter—until we cook [24].

5.4 Space Weather

One of the less obvious facts about interplanetary space is that it experiences weather. We briefly encountered the cause of this—the solar wind—in the previous chapter, when we were discussing the physics of solar sails. In that case, the solar wind proved to be a red herring; although it sounds like it ought to be the motive force behind a solar sail, it isn't. Instead, such sails are propelled by a much better known emanation from the the Sun, in the form of sunlight itself.

However, the solar wind does exist, and it's not a benign thing at all as far as space travellers are concerned. Consisting of a stream of high energy charged particles that are constantly being blasted out by the Sun, it constitutes a highly dangerous form of radiation. It's not electromagnetic radiation like the gamma rays produced by a nuclear bomb, but it can be just as dangerous to anyone unfortunate enough to get in its way.

The reason that most of us here on Earth are blissfully ignorant of the perils of the solar wind is that our planet has a built-in shield from it. This isn't provided by the Earth's atmosphere, as you might expect, but by its magnetic field. Because the solar particles carry an electric charge, their path is deflected when they encounter a magnetic field. In the case of the Earth's field, this has the fortunate effect of bending their trajectories around the planet. As a result, most of the harmful particles completely miss the Earth, which sits comfortably inside a protective "magnetosphere" (see Fig. 5.6).

Not all planets have magnetic fields, as Andy Weir explains in *Project Hail Mary*:

> Planets get magnetic fields if the conditions are right. You have to have a molten iron core, you have to be in the magnetic field of a star, and you have to be spinning. If all three of these things are true, you get a magnetic field. Earth has one—that's why compasses work [25].

Any planet that lacks a magnetic field, if it happens to be orbiting a star that produces a stellar wind similar to the Sun's, would be bathed in a constant stream of high-energy radiation. Needless to say, this would be very bad news

Fig. 5.6 A schematic illustration, not to scale, showing how the Earth's magnetic field deflects the potentially dangerous solar wind around the planet (NASA image)

for any life forms that tried to evolve there. In fact, more likely than not, such a planet would be totally lifeless.

In our case, on the other hand, the Earth's magnetosphere not only protects us down here on the surface of the planet, but also extends far enough into space to shield astronauts in Earth orbit, such as those on board the ISS. Beyond that, however, the radiation from the solar wind poses a potential hazard for any crews venturing further afield. For example, the Apollo astronauts who travelled to the Moon were exposed to a relatively high radiation dose, which could have been fatal if it had been kept up for an extended period of time. Fortunately, the Apollo missions were short enough that the crews were only exposed to a fiftieth of the fatal level of radiation.

For longer missions across interplanetary space, however, some form of additional protection will be needed. One possibility is to surround a spaceship with its own, artificially generated mini-magnetosphere—something that is actively being considered by researchers at the UK's Rutherford Appleton Laboratory. Their proposed system would use a powerful magnet similar to the ones used in hospital MRI scanners to create an effective "magnetic deflector shield" around the ship. If you think this sounds like something out of *Star Trek*, you're not the only one. It's a parallel that even the researchers themselves draw [26].

Even when they reach their destination, future astronauts may not be safe from cosmic radiation. Mars, for example, has a much weaker magnetic field than the Earth, and in consequence it's much more exposed to radiation. In fact, this is probably the biggest environmental hazard on the Red Planet, and the one that—in the real world—is most likely to have killed Mark Watney, as portrayed in either the book or film version of *The Martian*. Even Andy Weir himself has acknowledged that the lack of radiation protection is one of the story's most glaring implausibilities [27].

If radiation is going to be a hazard on Mars, then this is even more true in the airless environment of the Moon, with its extremely weak magnetic field. So any future lunar settlement will need to put serious thought into the matter of radiation protection. One possibility is to live deep below the surface, as described in Gregory Benford's 1977 novel *In the Ocean of Night*:

> More excavations were partially completed in the distance. Gradually a network of tubes was being punched by lasers, 10 metres beneath the shielding rock and dust. Set that deep, the quarters suffered little variation in temperature between lunar day and night and even the incessant rain of particles from the solar wind made radiation levels only slightly higher than those on Earth [28].

The effect of the unshielded solar wind on the Moon explains another minor mystery, which may have perplexed some people in the same way as the Apollo 11 flag. If there's no air on the Moon, why does the landscape look so smoothly weathered? Early SF illustrators, right up until the beginning of the 1960s, when the first close-up photographs of the lunar surface were sent back by robot probes, tended to portray it as consisting of jagged, sharp crags. This makes sense, given the Moon's lack of wind or flowing water, but it's not the way it actually looks at all (see Fig. 5.7). The truth is that the Moon is indeed subject to weathering, due to the steady erosion—over billions of years—caused by the solar wind particles raining down onto it.

It's fair to say that weathering on the Moon is a long-term process. The footprints left by the Apollo astronauts between 1969 and 1972, for example, are still as crisp and fresh as the day they were made. Eventually, however, the solar wind will take its toll on them. To quote Gregory Benford again, from *In the Ocean of Night*:

> The footprints Nikka made would, if left, survive for half a million years, until the fine spray of particles from the solar wind blurred them [29].

Although the solar wind is always there, its strength isn't constant. It becomes much more powerful during periods of enhanced activity on the Sun, such as solar flares and coronal mass ejections. The degree of solar activity tends to grow and wane on an 11-year cycle, although we're fortunate in

Fig. 5.7 The topography of the Moon shows clear signs of "weathering", due to the effects of the solar wind acting over billions of years (NASA image)

that—compared with certain other stars—the Sun is relatively consistent over time in the amount of energy it puts out.

Perhaps surprisingly, stars that are smaller than the Sun, such as red dwarfs, are generally much less well behaved. This is true, for example, of our next nearest star, the red dwarf Proxima Centauri. Like the Sun, Proxima occasionally produces flares, but they're far more violent, occasionally making the star thousands of times brighter than normal. As a NASA press release stated in 2021:

> The star known as Proxima Centauri, the Sun's nearest interstellar neighbour, turns out to have a hair-trigger temper—frequently erupting with potentially damaging stellar flares ... and these sizzling outbursts might be bad news for any potential lifeforms on the surface of a closely orbiting, probably rocky planet called Proxima b [30].

Elsewhere in the galaxy, some stars can be so unstable that they effectively explode. For observers on Earth, this can mean that a star that was too dim to see suddenly becomes extremely bright, appearing to be a brand new star—which gives the phenomenon its classical name of "nova". Today, astronomers use the term nova for a particular type of explosion occurring in certain binary star systems, while a "supernova" is the explosion accompanying the collapse of a very large star down to a black hole or neutron star (topics that were discussed earlier, in the chapter on gravity).

Our own Sun is neither part of a binary system nor massive enough to collapse to a neutron star or black hole, so there's zero chance of it ever "going nova". This rather dull fact, however, is no deterrent to SF writers who can't resist such an enticing story premise. As the *Encyclopedia of Science Fiction* tells us, "in numerous disaster stories the Sun goes nova" [31].

In fiction, however—even when it's written by scientifically literate authors—the term "going nova" is often used rather loosely to mean any extreme brightening of the Sun, far beyond what it normally experiences during its 11-year cycle. In James Blish's 1953 story "The Testament of Andros", for example, an astronomer believes (falsely, as it turns out) that the Sun is undergoing a cycle of ever-stronger flares that will eventually culminate in just such a disaster.

This maverick astronomer's colleagues are rightly sceptical of his claims. As one of them says, referring to a real-life astronomer named Robert Richardson, who wrote SF under the pseudonym of Philip Latham:

> You remind me of Dr Richardson's stories—you know, the ones he writes for those magazines, about the Sun going nova and all that.... But as a serious proposition it doesn't hold water. Our Sun just isn't the spectral type that goes nova; it hasn't ever even approached the critical instability percentage. It can't even produce a good flare of the [Proxima] Centauri-type [32].

A "nova"—in this sense of a vaguely defined explosion occurring on the Sun—is suspected to have occurred in Larry Niven's short story "Inconstant Moon", from 1971. The story's title comes from the fact that the Moon suddenly appears very much brighter on the night-time side of the Earth, so people there imagine that the Sun must have exploded while it was shining on the opposite hemisphere. Fortunately for the story's characters—who, like Niven himself, are based in the United States—the situation isn't as bad as they think. As the UK-based *Encyclopedia of Science Fiction* wryly observes, "the disaster proves to be a brief and just barely survivable (for America if not Europe) solar flare" [31].

Looking to the far distant future, as opposed to the very near future featured in Blish and Niven's stories, it's just conceivable that some major change in solar activity might lead to a civilization-destroying catastrophe. Arthur C. Clarke envisioned just such a possibility on a couple of occasions. For example, one of his earliest short stories, "Rescue Party" from 1946, is centred on the idea of humans escaping in spaceships from a Sun that's about to go nova [33].

Then, in his much later novel *The Songs of Distant Earth* (1986), Clarke revisits much the same idea—but with the benefit of several decades of further research by the astronomical community. In this case, he relates the world-ending disaster to what, at the time he wrote the novel, was a genuine mystery to astronomers—the so-called "solar neutrino problem". Without going into excessive technical detail, neutrinos are tiny subatomic particles that, according to astrophysical theory, should be created deep inside the Sun. From there, some of them will eventually find their way to Earth, where scientists could potentially detect them. However, as Clarke says in the novel:

> The experiment worked; solar neutrinos were detected. But there were far too few of them. There should have been 3 or 4 times as many as the massive instrumentation had succeeded in capturing [34].

As often happens when SF writers try to incorporate currently puzzling scientific discoveries into their fiction, Clarke's frightening assumption that the lack of solar neutrinos presaged a major disaster no longer holds water. Fortunately for us, the solar neutrino problem has now been solved, and while the solution isn't as exciting as the one that Clarke imagined, it's entirely non-catastrophic.

The topic of solar neutrinos does, however, raise a question that's all too easily overlooked in popular accounts of astronomy: how can scientists, who are confined to this one planet, know anything about the wider universe? It's pretty miraculous, when you think about it, that instruments here on Earth can tell us what's happening in the interior of the Sun, or in far-distant star systems scattered throughout the galaxy. This is a huge field in its own right, but it's worth taking a dip into it insofar as it relates to the subject of this book.

5.5 Viewing the Universe

In science fiction, too, there's a tendency to gloss over just how future space travellers are going to form a picture of their cosmic surroundings. The answer is that, more likely than not, they will use the same type of instruments currently employed by Earth-based astronomers to study objects thousands of light-years away. Whenever a character utters a glib remark that starts "sensors indicate…", they're almost certainly talking about some kind of telescope.

Of course, there's nothing science fictional about putting a telescope in space. Even for terrestrial astronomers, there are good reasons for doing this, even if it's only as far away as the low Earth orbit of NASA's Hubble space telescope. The most obvious reason is simply to get above the distorting effects of the Earth's atmosphere. Arthur C. Clarke made just this point in his non-fiction book *Report on Planet Three*, back in 1972. Describing what the space around Earth might look like in what, at that time, was the future, he wrote:

> Last, but perhaps most important of all, are the astronomical observatories with their vast, floating mirrors, scores of feet across, peering out across the billions of light-years and no longer half-blinded by the murk and haze of the atmosphere [35].

That phrase "murk and haze" refers not just to the masking effects of any pollution in the atmosphere, but also to the fact that even at night the sky glows slightly, never becoming completely black. This results from the scattering of light—whether it's coming from the stars or from terrestrial sources such as street lighting—by the air molecules bouncing around in the atmosphere.

There's another atmospheric effect, however, which is even more of a problem for astronomers. Everyone knows that stars "twinkle", rather than appearing as steady points of light, but this isn't actually an intrinsic property of the stars themselves. It's another consequence of the fact that the molecules in the air are always jiggling around, so that the ultra-thin rays of light coming from the stars appear to constantly change position. As a result, an Earthbound telescope can never produce an image that's anything like as sharp and clear as a space based one (see Fig. 5.8).

As an aside, there is a technique that astronomers can use today to make ground-based telescopes more effective than they used to be, by removing the "twinkling" effect mechanically. Called adaptive optics, this uses an advanced computer algorithm in conjunction with high-precision servo motors to constantly adjust the position of the telescopes mirrors in just such a way as to compensate for atmospheric distortion. It's a technique that—in the real world—is just coming into service in the newest generation of ground-based telescopes, but it was mentioned in a fictional context as long ago as 1988. Surprisingly, however, this wasn't a science fiction work by the likes of Arthur C. Clarke or Larry Niven, but Tom Clancy's Cold War techno-thriller *The Cardinal of the Kremlin*. Clancy is talking about an advanced laser weapon system rather than an astronomical telescope, but the basic technical principle is exactly the same:

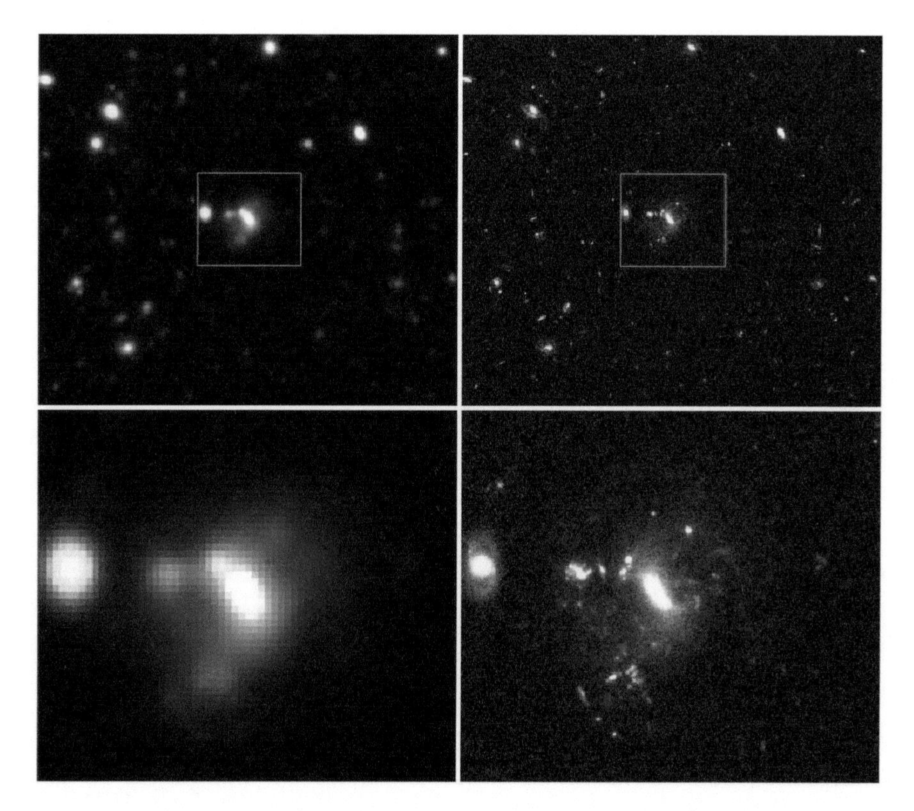

Fig. 5.8 Comparison of images taken with (left) an 8-metre diameter ground-based telescope and (right) the Hubble space telescope, which has a diameter of just 2.4 metres (NASA image)

> The mirror is the special part. It's composed of thousands of segments, and every segment is controlled by a piezoelectric chip. That's called "adaptive optics". We send an interrogation beam to the mirror … and get a reading on atmospheric distortion. The way the atmosphere bends the beam is analysed by computer. Then the mirror corrects for this distortion [36].

Although it's easy for us to forget, the visible light that we see with our eyes is only part of a much wider spectrum of electromagnetic waves. It's a point that Andy Weir makes in *Project Hail Mary*:

> Humans only see a tiny fraction of all the wavelengths of light out there. We evolved to see the wavelengths that are most plentiful on Earth [37].

Aliens on another planet, where the physical conditions are different from ours, might evolve different sensory mechanisms. Arthur C. Clarke imagined one such possibility in his 1971 novella "A Meeting with Medusa", set in the thick, high-pressure atmosphere of Jupiter. The eponymous medusa, a creature living inside this atmosphere, has evolved biological sensors that work at wavelengths of around a metre—comparable to FM radio broadcasts on Earth. As a character in the story observes:

> That's something nature never got round to on Earth … nothing ever developed a radio sense. Why bother where there was so much light? But it's different here. Jupiter is drenched in radio energy [38].

Although humans may not have naturally evolved radio sensors, we do, of course, have radio technology—and astronomers can employ radio telescopes to observe aspects of the universe that may not be detectable by their optical counterparts. One such instrument, supposedly "the biggest radio telescope … anywhere" at the time the story takes place, is featured in the 1962 novel *A for Andromeda*, written by astronomer Fred Hoyle in collaboration with John Elliot. Although their fictional telescope is based in Britain, its description sounds rather like the giant Arecibo telescope in Puerto Rico (which was built the following year, 1963, and operated until 2020):

> It stood in front of them: three huge pillars curving together at the top to form a triangular arch, dark and stark against the ebbing sky. Hollowed out of the ground between the uprights lay a concrete bowl the size of a sports arena, and above, suspended from the top of the arch, a smaller metal bowl looked downwards and pointed a long antenna at the ground.

The operation of the telescope is described by one of the technicians in the story as follows:

> You know the principle of the thing? Any radio emission from the sky strikes the bowl and is reflected to the aerial, and received and recorded on the equipment.… This bank of computers works out the azimuth and elevation of whatever source you want to focus on and keeps it following [39].

Most people know that radio telescopes tend to be enormous, but why do they have to be so big? The answer is that the resolving power of a telescope—the smallest detail it can make out—is proportional to the wavelength it operates at divided by its diameter. Since we want this to be a small number, and since the wavelength of radio waves is so much larger than that of visible light, we consequently need a much greater diameter. As a result, each generation of radio telescopes tends to be larger than its predecessor—a fact that works in favour of the protagonist of James Blish's story "The Testament of Andros". As we saw earlier, this character is chiefly interested in the physics of the Sun—which, being such a large object in the sky, doesn't need the same kind of resolution as a distant star would. This makes him a winner when an old telescope is replaced by a newer, larger one:

> Completion of the 1,000-inch freed the 600-inch paraboloid antenna for my use in solar work. The smaller instrument had insufficient beam width between half-power points for the critical stellar studies, but it was more suitable for my purpose [40].

Returning to Fred Hoyle, another use of radio telescopes is seen in his famous novel *The Black Cloud*, from 1957. As we've seen in previous chapters, this deals with an interstellar cloud that's heading towards the Solar System, and at one point in the story a group of astronomers are discussing how they can work out its speed of approach. As one of them says:

> The Australian radio boys could get the information for us. It's very likely that the cloud consists mainly of hydrogen, and it should be possible to get a Doppler shift on the 21 centimetre line [41].

As so often with Hoyle, this packs a lot of detailed physics into a relatively terse statement. Basically, there are two principles involved here. The first is one of the oldest applications of radio astronomy, to the detection of neutral hydrogen gas in the galaxy. This happens to emit a characteristic signal, known as a "spectral line", at a wavelength of 21 cm. For astronomers, it's very fortunate that it does, because it allows them to map the distribution of this gas—which would otherwise be completely invisible—simply by looking out for its distinctive 21-cm emissions.

The other principle that Hoyle is drawing on is the Doppler shift, whereby the wavelength of a signal is altered if the source is approaching or receding from the observer. In an astronomical context, its most familiar consequence is the so-called "redshift" of distant galaxies that are moving away from us. This means that the light coming from them is stretched out, and hence shifted towards the red—or long wavelength—end of the spectrum. On the other hand, with Hoyle's fast-approaching cloud, the 21-cm emissions would be squashed up as they come towards us, and hence appear at a wavelength slightly smaller than 21 cm. By measuring the size of the shift, an observer can work out what the cloud's speed must be.

The 21-cm line that looms large in radio astronomy is really just one aspect of a much wider subject called spectroscopy. If you've ever wondered how astronomers can tell you what the chemical composition of a distant star or nebula is, then spectroscopy is the answer. All chemical elements and compounds produce characteristic signals analogous to the 21-cm line, though these are more commonly found in and around the visible part of the electromagnetic spectrum. The principle is explained, in characteristically easy-to-understand language, by Andy Weir in *Project Hail Mary*:

> When light hits gas molecules, the electrons get all worked up. Then they calm down and re-emit the energy as light. But the frequency of the photons they emit is very specific to the molecules involved. Astronomers used this for decades to know what gases are out there far, far away. That's what spectroscopy is all about [42].

While many outsiders imagine that astronomers spend all their time taking photographic images of celestial objects, in reality they use spectroscopy just as much, if not more—and have done for just as long as they've been taking astronomical photographs. It's a standard procedure that Arthur C. Clarke describes in his 1955 novel *Earthlight*, when astronomers at a fictional observatory on the Moon discover a supernova. When one of them rushes to analyse it, he's not interested in what it looks like, but in what its spectrum is:

> Had he looked into the beam, the sheer glare of Nova Draconis would have blinded him—and, as compared with his instruments, his eyes could tell him practically nothing. He switched the electronic spectrometer into place, and started it scanning. It would explore the spectrum of Nova Draconis with patient accuracy, working down through yellow, green, blue into the violet and far ultra-violet, utterly beyond the range of the human eye. As it scanned it would trace on moving tape the intensity of every spectral line [43].

Spectral lines are easiest to detect in hot gas, which makes them very useful in the case of stars and nebulae, but less so with a cooler, solid object like an asteroid. In this case, it helps to have a space probe to stir things up a bit—as Clarke showed in *2001: A Space Odyssey* (the book, not the film). The astronauts on board spaceship *Discovery* make an asteroid more suitable for spectral analysis by firing a small probe into it as they pass:

> The probe contained no instruments; none could survive a collision at such cosmic speeds. It was merely a small slug of metal, shot out from *Discovery* on a course which should intersect that of the asteroid.

The rationale for this becomes clearer after the probe hits its target:

> Against the darkened portion of the asteroid there was a sudden, dazzling explosion of light. The tiny slug had impacted at meteoric speed; in a fraction of a second, all its energy had been transformed into heat. A puff of incandescent gas had erupted briefly into space; aboard *Discovery* the cameras were recording the rapidly fading spectral lines. Back on Earth, experts would analyse them, looking for the tell-tale signatures of glowing atoms. And so, for the first time, the composition of an asteroid's crust would be determined [44].

A very similar experiment took place in the real world—this time with a comet rather than an asteroid—during NASA's Deep Impact mission in 2005. Just like the fictional *Discovery*, the spacecraft fired a small probe into comet Tempel 1 and observed the results with an array of spectrometers and other instruments (see Fig. 5.9).

Fig. 5.9 When NASA's Deep Impact spacecraft fired a small test probe at comet Tempel 1, it was reminiscent of a similar scene in Arthur C. Clarke's novel *2001: A Space Odyssey* (NASA image)

5.6 What Have we Learned?

In the course of this book, we have explored a large part of what might be called the "everyday" physics of space travel. In doing so, we have seen innumerable instances of this physics being incorporated naturally—and accurately—in science fiction novels and short stories. This is a great antidote to the common perception, based on mass-market Hollywood movies and TV shows, that SF always gets its physics wrong. Of course, not all SF is scientifically accurate—probably only a small minority of it—but when it is, it can add enormously to the pleasure of reading it.

Authors who have cropped up in this context range from Jules Verne in the nineteenth century to ex-astronaut Chris Hadfield in the 21st, and include Isaac Asimov, Robert Heinlein, Hal Clement, Fred Hoyle, Stephen Baxter and Gregory Benford. The three names that have been mentioned most frequently, however, are those of Arthur C. Clarke, Andy Weir and Larry Niven. All three have taken particular pleasure, not just in getting their physics correct, but in making it central to the plots of many of their stories. To repeat a quote from Larry Niven that appeared in the first chapter of this book, space is "a place to learn physics where you can watch it happen" [45].

At first sight, this may seem an odd thing to say; surely physics works in just the same way here on Earth? It's true that it does, but observing it in action can be obfuscated by the constant presence of the Earth's atmosphere and gravitational field. This means that, once you get free of such things, the physics of outer space can seem counter-intuitive in many different ways. It's almost certainly for this reason that Hollywood movie producers prefer their own brand of physics to that of the real world.

Even so, it can be insulting to a viewer's intelligence to portray, for example, the sound of an explosion in space, or a spaceship banking like an aircraft in order to change direction, or slowing to a stop as soon as its engines are turned off. The authentic physics of how such things really work is not only well established, but it has been known to scientists for centuries. In fact, the same is true of virtually all the physics covered in this book. Here's a quick rundown of some of the areas discussed:

- The first chapter addressed some of the most basic physics, of the kind that many people encounter in school, such as Archimedes' principle—which dates from as far back as the third century BCE—and the motion of a pendulum, as well as the conservation of energy and momentum, which are arguably the most universal physical principles of all;

- The second chapter focused on the subject of gravity—both the basic theory developed by Sir Isaac Newton in the seventeenth century, as well as a number of implications arising from it that occasionally feature in the better kind of science fiction, such as tidal forces and Lagrange points;
- The third chapter looked at the field of orbital dynamics, which is far more central to spaceflight than many people realize—essentially governing all spacecraft trajectories within the Solar System, not just those that happen to be "in orbit" around the Earth or another planet;
- The fourth chapter was titled "Rocket Science", but actually ranged more broadly than that—dealing not only with the basic physics of rocketry but also other scientifically feasible modes of spaceflight, from ion drives and nuclear propulsion to light sails and Bussard ramjets;
- Finally, this chapter has considered a number of physical processes occurring beyond the Earth's atmosphere, from vacuum physics and spacecraft life support systems to space weather and the exploitation of electromagnetic waves to gather information on astronomical objects using telescopes and other instruments.

It really can't be over-stressed just how basic most of this material is, within the overall context of currently understood physics. With the exception of a few items such as relativistic time dilation and neutron stars, which date from the early twentieth century, it was all well known stuff by the middle of the nineteenth century. In spite of this, it's essentially all the physics we require as far as the kind of realistic science fiction we've been talking about is concerned.

Significantly, at no point in this book have we felt the need to invoke the dreaded Q-word. Although it's now a well-established part of modern physics, this is often bandied about by the lazier kind of SF authors to mean that, in effect, anything they care to imagine is scientifically possible. Or, to give the last word to the *TV Tropes* website:

> A well-known narrative device is that Quantum Mechanics Can Do Anything. Time-travel? God-mode superpowers? Death rays? Yes, all of that. Expect complex equations, a lot of technobabble and explanations of the Schrödinger's cat experiment, and the observer effect that make every conscious being a potential reality warper. It's a lot easier when you don't have to do any research [46].

References

1. A.C. Clarke, *The Sands of Mars* (Pan Books, London, 1964), p. 13
2. L. Niven, Wrong Way Street, in *Convergent Series*, (Orbit Books, London, 1986), pp. 224–225
3. A. Weir, *Project Hail Mary* (Kindle Edition), loc. 5917
4. I. Asimov, *The Relativity of Wrong* (Oxford University Press, Oxford, 1989), p. 208
5. C. Hadfield, *The Apollo Murders* (Kindle Edition), loc. 5928
6. G.O. Smith, *The Complete Venus Equilateral* (Ballantine Books, New York, 1976), p. 35
7. C. Hadfield, *The Apollo Murders* (Kindle Edition), loc. 1800
8. A. Weir, *Project Hail Mary* (Kindle Edition), loc. 2237
9. T. McMullan, Was the Hole on the International Space Station Really Sabotage?, https://www.wired.co.uk/article/international-space-station-leak-hole-drill-what-happened
10. A.C. Clarke, *The Sands of Mars* (Pan Books, London, 1964), pp. 53–59
11. A. Weir, *Project Hail Mary* (Kindle Edition), loc. 6929
12. A. May, *The Science Behind Jules Verne's Moon Novels* (Post-Fortean Books, 2018), pp. 65–66
13. A. Weir, *The Martian* (Del Rey, New York, 2014), p. 5
14. R.A. Heinlein, Misfit, in *Minds Unleashed*, (Grosset & Dunlap, New York, 1970), pp. 171–173
15. A.C. Clarke, *The Other Side of the Sky* (Signet Books, New York, 1959), p. 34
16. A.C. Clarke, Silence Please, in *Tales from the White Hart*, (Sidgwick & Jackson, London, 1972), p. 14
17. A.C. Clarke, *Earthlight* (Pan Books, London, 1966), p. 52
18. R.A. Heinlein, *Rocketship Galileo* (New English Library, London, 1971), p. 115
19. A.C. Clarke, *A Fall of Moondust* (Pan Books, London, 1971), pp. 202–203
20. A.C. Clarke, *A Fall of Moondust* (Pan Books, London, 1971), p. 29
21. L. Niven, *A World out of Time* (Orbit Books, London, 1988), p. 28
22. A. May, *The Science Behind Jules Verne's Moon Novels* (Post-Fortean Books, 2018), pp. 69–70
23. A. Weir, *The Martian* (Del Rey, New York, 2014), p. 72
24. A.C. Clarke, *A Fall of Moondust* (Pan Books, London, 1971), p. 37
25. A. Weir, *Project Hail Mary* (Kindle Edition), loc. 3808
26. Mini magnetospheres, Rutherford Appleton Laboratory, https://www.ralspace.stfc.ac.uk/Pages/Mini-magnetospheres.aspx
27. C. Moskowitz, "Martian Astronaut Would Get Cancer if Mission Were Real, Author Says", https://www.scientificamerican.com/article/martian-astronaut-would-get-cancer-if-mission-were-real-author-says1/
28. G. Benford, *In the Ocean of Night* (Kindle Edition), loc. 4009
29. G. Benford, *In the Ocean of Night* (Kindle Edition), loc. 5057

30. P. Brennan, Neighbouring Star's Bad Behaviour, https://exoplanets.nasa.gov/news/1680/neighboring-stars-bad-behavior-large-and-frequent-flares/
31. Sun, *Encyclopedia of Science Fiction*, https://sf-encyclopedia.com/entry/sun
32. J. Blish, The Testament of Andros, in *Alpha 1*, (Ballantine Books, London, 1971), p. 52
33. A.C. Clarke, Rescue Party, in *The Sentinel*, (Berkley Books, New York, 1986), pp. 13–38
34. A.C. Clarke, *The Songs of Distant Earth* (Harper-Collins, London, 1994), p. 13
35. A.C. Clarke, *Report on Planet Three*, vol 63 (Corgi Books, London, 1973), pp. 63–531
36. T. Clancy, *The Cardinal of the Kremlin* (Harper Collins, London, 1994), p. 61
37. A. Weir, *Project Hail Mary* (Kindle Edition), loc. 2875
38. A.C. Clarke, A Meeting with Medusa, in *The Sentinel*, (Berkley Books, New York, 1986), p. 244
39. F. Hoyle, J. Elliot, *A for Andromeda* (Corgi Books, London, 1972) pp. 7, 9, 12–3
40. J. Blish, The Testament of Andros, in *Alpha 1*, (Ballantine Books, London, 1971), p. 50
41. F. Hoyle, *The Black Cloud* (Penguin Books, London, 1960), p. 44
42. A. Weir, *Project Hail Mary* (Kindle Edition), loc. 1482
43. A.C. Clarke, *Earthlight* (Pan Books, London, 1966), pp. 23–24
44. A.C. Clarke, *A Space Odyssey*, vol 1968 (Arrow Books, London, 2001), p. 120
45. L. Niven, At the Bottom of a Hole, in *Tales of Known Space*, (Ballantine Books, New York, 1975), p. 99
46. Quantum Mechanics Can Do Anything, *TV Tropes*, https://tvtropes.org/pmwiki/pmwiki.php/Main/QuantumMechanicsCanDoAnything

Printed in the United States
by Baker & Taylor Publisher Services